# Toxin-Host Interaction of Clostridium Toxins

# Toxin-Host Interaction of Clostridium Toxins

Guest Editor

**Jonatan Dorca-Arévalo**

MDPI

Basel • Beijing • Wuhan • Barcelona • Belgrade • Novi Sad • Cluj • Manchester

*Guest Editor*
Jonatan Dorca-Arévalo
Department of Pathology and
Experimental Therapeutics
University of Barcelona
Barcelona
Spain

*Editorial Office*
MDPI AG
Grosspeteranlage 5
4052 Basel, Switzerland

This is a reprint of the Special Issue, published open access by the journal *Toxins* (ISSN 2072-6651), freely accessible at: https://www.mdpi.com/journal/toxins/special_issues/SZ265YQGNR.

For citation purposes, cite each article independently as indicated on the article page online and as indicated below:

| Lastname, A.A.; Lastname, B.B. Article Title. *Journal Name* **Year**, *Volume Number*, Page Range. |
|---|

**ISBN 978-3-7258-7080-6 (Hbk)**
**ISBN 978-3-7258-7081-3 (PDF)**
**https://doi.org/10.3390/books978-3-7258-7081-3**

Cover image courtesy of Jonatan Dorca-Arévalo

# Contents

# About the Editor

**Jonatan Dorca-Arévalo**

Jonatan Dorca-Arévalo is an associate professor at the Department of Pathology and Experimental Therapeutics of the University of Barcelona. His research focuses on the effects of *Clostridium perfringens* epsilon toxin on the nervous and renal systems, investigating its molecular mechanisms. He is actively studying the relationship between epsilon toxin and multiple sclerosis, aiming to understand its contribution to disease pathology. He also explores the diagnostic and therapeutic potential of epsilon toxin as a tool to target MAL, which is highly expressed in aggressive endometrial tumors, providing a potential approach for both detection and treatment. Jonatan obtained his PhD at the University of Barcelona, Campus Bellvitge, where he studied the binding and action of epsilon toxin. He conducted research on spinal muscular atrophy in the Molecular Bases of Diseases Research Group at the Sant Pau Research Institute (IIB Sant Pau) under the coordination of Pablo Fuentes Prior, gaining expertise in protein expression and X ray crystallography. As a postdoctoral researcher in the Department of Cell Biology, Physiology, and Immunology at the Faculty of Biology of the University of Barcelona, in the laboratory of Eduardo Soriano, he studied the septohippocampal GABAergic pathway in Alzheimer's disease, characterizing Tau hyperphosphorylation, synaptic alterations, and their impact on neuronal function. He has also investigated the role of epsilon toxin in multiple sclerosis and its contribution to neuroinflammation through extracellular vesicles at IDIBELL. His research interests include clostridial toxin biology, neurotoxicity, vesicle-mediated toxin transport, and neurodegenerative diseases. He collaborates on projects such as the development of a vaccine candidate targeting ETX and multiple sclerosis with One Health Ventures Ltd. He has published 16 scientific papers, presented at over 30 conferences, and serves as the Guest Editor of a Special Issue in *Toxins*.

toxins

MDPI

*Editorial*

# Toxin–Host Interaction of *Clostridium* Toxins

**Jonatan Dorca-Arévalo [1,2,3]**

1 Department of Pathology and Experimental Therapeutics, Faculty of Medicine and Health Sciences-Campus Bellvitge, University of Barcelona, 08907 Barcelona, Spain; jondorca@ub.edu
2 Laboratory of Molecular and Cellular Neurobiology, Bellvitge Biomedical Research Institute (IDIBELL), 08907 L'Hospitalet de Llobregat, Spain
3 Institute of Neuroscience, Bellvitge Health Sciences Campus, University of Barcelona, Carrer de la Feixa Llarga, s/n, 08907 L'Hospitalet de Llobregat, Spain

Clostridial toxins, involved in a wide range of serious diseases, affect humans and animals. Among the best known are the neurotoxins produced by *Clostridium botulinum* and *Clostridium tetani*, which are considered to be one of the most potent bacterial toxins known, with botulinum toxin being the most potent biological toxin identified to date. In addition, many clostridial toxins cause gastrointestinal disorders that can lead to severe health complications and, in some cases, death. In this Special Issue (SI), "Toxin–Host Interaction of *Clostridium* Toxins", we present recent research on toxins produced by *Clostridioides difficile*, the most significant species due to its clinical impact and high prevalence, particularly in hospital settings, and *Clostridium perfringens*, which causes severe infections such as gas gangrene, a rapidly fatal condition, as well as enterotoxemia.

The toxins produced by *Clostridioides difficile*, TcdA and TcdB, possess glucosyltransferase activity and cause severe symptoms ranging from diarrhea to pseudomembranous colitis. Both toxins are internalized into host cells via receptor-mediated endocytosis. Their enzyme domains glucosylate small GTPases, thereby disrupting actin cytoskeleton regulation. Interestingly, pharmacological inhibition of Hsp70 using the inhibitor VER-155008, as well as the antiemetic drug domperidone, was shown to protect cells from TcdB intoxication. Domperidone prevented the membrane translocation of TcdB's glucosyltransferase domain into the cytosol and also conferred protection against TcdA and the binary CDT toxin produced by hypervirulent *C. difficile* strains [1]. These findings identify Hsp70 as a promising therapeutic target against *Clostridioides difficile*.

Regarding *Clostridium perfringens*, it is a spore-forming, Gram-positive anaerobic pathogen that causes various disorders in humans and animals. Recently, a multidrug-resistant strain (IRMC2505A) was isolated from the fecal sample of a patient clinically suspected of a gastrointestinal infection, who had a recent history of antibiotic exposure and diarrhea. Nineteen antimicrobial resistance genes and virulence factors were identified in the IRMC2505A strain through 16S rRNA sequencing, followed by whole-genome analysis and bioinformatics mapping against the CARD and VFDB databases. This is the first whole-genome report of this strain from Saudi Arabia and highlights the need to understand its resistance and virulence profiles to guide control strategies [2].

The most lethal toxin produced by *Clostridium perfringens* is the epsilon toxin (ETX). This toxin, produced by types B and D, causes generalized edema and is capable of crossing the blood–brain barrier, resulting in neurotoxicity that can lead to animal death. Interestingly, ETX has recently been linked to a potential association with multiple sclerosis in humans, although further studies are required to confirm this relationship. While ruminants are the natural hosts for *C. perfringens* types B and D and, consequently, for ETX-mediated neurotoxicity, laboratory mouse models have been developed that exhibit similar toxic effects, allowing experimental investigation of the toxin's mechanisms. ETX is classified as a member of the pore-forming toxin (PFT) family due to its ability to form pores in the plasma membrane of sensitive cell types such as MDCK, 1G11, FRT, and MOLT4 cells. Despite this classification, the specific molecular changes induced in affected cells that ultimately lead to cell death remain poorly understood. To address this knowledge gap, a recent study employed quantitative proteomics and phosphoproteomics, utilizing TMT labeling and LC–MS/MS, to systematically analyze changes in protein expression and phosphorylation in MDCK cells exposed to ETX. This approach revealed a large number of differentially expressed and phosphorylated proteins, many of which were enriched in pathways related to RNA processing and the spliceosome, implicating these processes as central components of ETX-induced cellular toxicity. Functional validation experiments using pharmacological inhibitors and RNA interference further demonstrated that key proteins, including SRSF1, SF3B2, THOC2, and SRPK1, which are involved in spliceosome-related pathways, play a critical role in mediating ETX-induced cell death. These findings suggest that disruption of mRNA splicing is a major mechanism underlying ETX toxicity and provide a foundation for future research aimed at identifying therapeutic targets to mitigate the effects of this potent neurotoxin [3].

The Myelin and Lymphocyte (MAL) protein is the most widely accepted putative receptor and is primarily expressed in myelin and in the plasma membrane of T-lymphocytes. In this context, an interesting study investigating the cytotoxic effects of ETX on primary human lymphocytes demonstrated that ETX preferentially binds to and kills these cells in a time- and dose-dependent manner. Using flow cytometry, the authors quantified ETX binding and cytotoxicity across lymphocyte subsets ($CD4^+$, $CD8^+$, and $CD19^+$), finding the highest binding and killing in $CD4^+$ cells. To determine whether receptor expression accounted for this cell-type preference, the study measured MAL gene expression by RT-qPCR, showing that $CD4^+$ lymphocytes express higher levels of MAL. These results indicate that primary human lymphocytes are susceptible to ETX and support the hypothesis that MAL is a key receptor mediating toxin binding. The findings suggest that ETX may influence human immune responses and potentially implicate immune cells in conditions like multiple sclerosis [4]. These results align with the results obtained by Dorca-Arévalo's laboratory using MOLT4 cells. In addition, the generation of extracellular vesicles (EVs) for the affected cells is induced by ETX, and this action is dependent on MAL-GFP expression in the plasma membrane, as demonstrated by the HeLa cell line. Moreover, in those EVs, the presence of ETX and MAL was detected, suggesting that host cells may shed EVs in response to pore formation, as other PFTs can do [5]. This finding provides a new tool to study the mechanism of ETX action and offers insight into how ETX interacts with its receptor.

Vaccines play a critical role in preventing *Clostridium* infections, and there is currently intense research focused on developing new, more efficient strategies to improve safety, immunogenicity, and coverage. In line with these efforts, a recent review traces the evolution of vaccine design from early live attenuated and inactivated vaccines to modern approaches, including recombinant protein and subunit vaccines targeting bacterial toxins. It highlights advances such as genetically detoxified proteins, mRNA platforms, and other novel technologies, emphasizing how a deeper understanding of pathogen–host interactions can guide the development of safer and more effective toxin-based vaccines [6]. Building on this, a recent study developed a dissolving microneedle (dMN) vaccine for *Clostridium perfringens* epsilon toxin (ETX) by formulating recombinant ETX protein with poly(lactic-co-glycolic acid) (PLGA) nanoparticles modified with dimethyldioctadecylammonium bromide (DDAB) to enhance antigen adsorption and immune response. PLGA nanoparticles were prepared and integrated into a fish gelatin microneedle matrix that

 https://doi.org/10.3390/toxins18020067

could successfully penetrate skin. In vivo immunization via the dMN patch and subcutaneous injections revealed that both routes induced similar antibody titers, but the dMN vaccine provided superior protection in neutralization assays against lethal ETX challenge.

In vivo neutralization experiments showed 100% survival in the dMN group challenged with 100 × $LD_{50}$ ETX, while the aluminum-adjuvant group showed lower survival, demonstrating the improved protective efficacy of the PLGA-based microneedle vaccine [7].

These promising experimental approaches complement broader research efforts, as highlighted in a recent review summarizing the epidemiology, toxin biology, and vaccine development strategies for *Clostridium perfringens*. The review provides a comprehensive overview of current knowledge on the pathogenesis of *C. perfringens* toxins, their molecular mechanisms, and ongoing attempts to develop effective immunization strategies, emphasizing both traditional toxoid vaccines and novel platforms such as recombinant and nanoparticle-based vaccines [8]. Importantly, the review does not identify a single superior vaccine; instead highlights the ongoing need for research to develop safer and more effective immunization approaches.

**Funding:** This research received no external funding.

**Acknowledgments:** As Guest Editor, I thank the authors who contributed to the success of this SI of Toxins, as well as to the MDPI team for their valuable editorial support.

**Conflicts of Interest:** The author declares no conflicts of interest.

## References

1. Braune-Yan, M.; Jia, J.; Wahba, M.; Schmid, J.; Papatheodorou, P.; Barth, H.; Ernst, K. Domperidone Protects Cells from Intoxication with Clostridioides difficile Toxins by Inhibiting Hsp70-Assisted Membrane Translocation. *Toxins* **2023**, *15*, 384. [CrossRef] [PubMed]
2. AlJindan, R.; AlEraky, D.M.; Farhat, M.; Almandil, N.B.; AbdulAzeez, S.; Borgio, J.F. Genomic Insights into Virulence Factors and Multi-Drug Resistance in. *Toxins* **2023**, *15*, 359. [CrossRef] [PubMed]
3. Yue, N.; Huang, J.; Dong, M.; Li, J.; Gao, S.; Wang, J.; Wang, Y.; Li, D.; Luo, X.; Liu, T.; et al. Proteome and Phosphoproteome Profiling Reveal the Toxic Mechanism of. *Toxins* **2024**, *16*, 394. [CrossRef]
4. Shetty, S.V.; Mazzucco, M.R.; Winokur, P.; Haigh, S.V.; Rumah, K.R.; Fischetti, V.A.; Vartanian, T.; Linden, J.R. Epsilon Toxin Binds to and Kills Primary Human Lymphocytes. *Toxins* **2023**, *15*, 423. [CrossRef] [PubMed]
5. Dorca-Arévalo, J.; Santana-Ruiz, A.; Torrejón-Escribano, B.; Martín-Satué, M.; Blasi, J. Epsilon Toxin from *Clostridium perfringens* induces the generation of extracellular vesicles in HeLa cells overexpressing Myelin and Lymphocyte protein. *Toxins* **2024**, *16*, 525. [CrossRef]
6. Gupta, S.; Pellett, S. Recent Developments in Vaccine Design: From Live Vaccines to Recombinant Toxin Vaccines. *Toxins* **2023**, *15*, 563. [CrossRef] [PubMed]
7. Wan, W.; Li, Y.; Wang, J.; Jin, Z.; Xin, W.; Kang, L.; Li, X.; Cao, Y.; Yang, H.; Gao, S. PLGA Nanoparticle-Based Dissolving Microneedle Vaccine of. *Toxins* **2023**, *15*, 461. [CrossRef] [PubMed]
8. Titball, R.W. The Molecular Architecture and Mode of Action of Clostridium perfringens ε-Toxin. *Toxins* **2024**, *16*, 180. [CrossRef] [PubMed]



 https://doi.org/10.3390/toxins18020067

toxins

MDPI

*Article*

# Genomic Insights into Virulence Factors and Multi-Drug Resistance in *Clostridium perfringens* IRMC2505A

**Reem AlJindan [1], Doaa M. AlEraky [2], Maha Farhat [3], Noor B. Almandil [4,*], Sayed AbdulAzeez [5] and Jesu Francis Borgio [5]**

1 Department of Microbiology, College of Medicine, Imam Abdulrahman Bin Faisal University, Dammam 31441, Saudi Arabia; raljindan@iau.edu.sa
2 Department of Biomedical Dental Science, Microbiology and Immunology Division, Collage of Dentistry, Dammam 31441, Saudi Arabia; dmaleraky@iau.edu.sa
3 Department of Biochemistry, College of Medicine, Imam Abdulrahman Bin Faisal University, Dammam 31441, Saudi Arabia; mfarhat@iau.edu.sa
4 Department of Clinical Pharmacy Research, Institute for Research and Medical Consultations (IRMC), Imam Abdulrahman Bin Faisal University, Dammam 31441, Saudi Arabia
5 Department of Genetic Research, Institute for Research and Medical Consultations (IRMC), Imam Abdulrahman Bin Faisal University, Dammam 31441, Saudi Arabia; asayed@iau.edu.sa (S.A.); fbalexander@iau.edu.sa (J.F.B.)
* Correspondence: nbalmandil@iau.edu.sa

**Abstract:** *Clostridium perfringens* is a spore-forming, Gram-positive anaerobic pathogen that causes several disorders in humans and animals. A multidrug-resistant *Clostridium* strain was isolated from the fecal sample of a patient who was clinically suspected of gastrointestinal infection and had a recent history of antibiotic exposure and diarrhea. The strain was identified by 16s rRNA sequencing as *Clostridium perfringens*. The strain's pathogenesis was analyzed through its complete genome, specifically antimicrobial resistance-related genes. The *Clostridium perfringens* IRMC2505A genome contains 19 (*Alr, Ddl, dxr, EF-G, EF-Tu, folA, Dfr, folP, gyrA, gyrB, Iso-tRNA, kasA, MurA, rho, rpoB, rpoC, S10p,* and *S12p*) antibiotic-susceptible genetic species according to the k-mer-based detection of antimicrobial resistance genes. Genome mapping using CARD and VFDB databases revealed significant ($p$-value = $1 \times 10^{-26}$) genes with aligned reads against antibiotic-resistant genes or virulence factors, including phospholipase C, perfringolysin O, collagenase, hyaluronidase, alpha-clostripain, exo-alpha-sialidase, and sialidase activity. In conclusion, this is the first report on *C. perfringens* from Saudi Arabia that conducted whole genome sequencing of IRMC2505A and confirmed the strain as an MDR bacterium with several virulence factors. Developing control strategies requires a detailed understanding of the epidemiology of *C. perfringens*, its virulence factors, and regional antimicrobial resistance patterns.

**Keywords:** *Clostridium*; anaerobic pathogen; multi-drug resistant; genome mapping; resistance genes; virulence

**Key Contribution:** Whole genome analysis of *C. perfringens* showed antibiotic-resistant genes associated with multi-drug resistance. This first-of-its-kind WGS study reported *C. perfringens* from Arabian Peninsula.

## 1. Introduction

*Clostridium perfringens* (*C. perfringens*) is a spore-forming, Gram-positive bacterium. *C. perfringens* was first discovered in 1981 at the Johns Hopkins Hospital. It is a causative agent of gastroenteritis, myonecrosis, and enterotoxemia. Initially, it was known as *Bacillus aerogenes capsulatus*. Before the final name, *Clostridium perfringens*, this pathogen was called *Bacillus welchii* [1–10]. *C. perfringens* can produce spores that are tolerant to environmental stress and have a wide variety of toxins involved in the pathogenesis of diseases. Spores of this pathogen can tolerate cooking temperatures and survive. As a result, *C. perfringens* can grow in foods that have been stored incorrectly. The most frequent causes of outbreaks are associated with incorrectly prepared or reheated food materials such as meats, poultry, or gravy and improper management of patients with *C. perfringens* infections before the exact diagnosis of causative agents [10]. There is evidence that type A and type C toxins can harm humans. Most cases of non-foodborne diarrheal illness and food poisoning linked to *C. perfringens* are caused by type A toxin. The pathophysiology of this pathogen is mostly associated with tissue necrosis caused by the toxin of *C. perfringens*. The majority of toxins from *C. perfringens* have pore-forming behaviors that open up, allowing water to flood in and cause swelling followed by cell death. The fermentation of glucose in this pathogen results in the generation of histotoxic gas, which is a defining feature of *C. perfringens* infection [2,5,10]. This anaerobic pathogen inhabits diverse environments and causes several disorders in humans and animals, including histotoxic (gas gangrene) infections and intestinal-associated disorders. Its pathogenicity depends on its massive production of more than 20 toxins, which varies according to different strains of *C. perfringens* [1–3]. These strains are classified as A–G toxinotypes according to major extracellular toxins (α, β, ε, and ι toxins). Previous comparative genomic studies on this gut pathogen indicated considerable genome variation [4]. The pathogenesis of intestinal-related diseases is characterized by rapid anaerobic growth in host tissues accompanied by the in vivo production of several key pore-forming toxins, including α-, β-, and β2-toxins, which disrupt epithelial barrier function and induce tissue necrosis or histotoxic infection [4].

Phospholipase C (plc) is the main virulent factor in *C. perfringens*-associated diseases. It is a chromosomally encoded α-toxin produced by all strains of this pathogen [5]. *C. perfringens* enterotoxin (*cpe*) is a clinically important enterotoxin gene in a few strains. This enterotoxin is the etiology of most gastrointestinal-associated diseases, including antibiotic-associated diarrhea, food-poisoning outbreaks, and nosocomial diarrheal disease. Outbreaks caused by *C. perfringens* are common, with severe illness or death reported. Current public health interventions reduce contamination at the food production source and cross-contamination in the food preparation environment [6–8]. Different literature has reported the characteristic resistance of *C. difficile* and its genotype profile using the resolution tool provided by whole genome sequencing (WGS) to analyze genetic diversity and assess *C. difficile* strains associated with infection recurrences and hospital outbreaks. Moreover, novel strains can spread quickly across continents and nations, negatively affecting global health. Some strains are more contagious than others for various reasons, including sporulation, antibiotic resistance, and toxin production. Using a unified typing strategy such as WGS will dramatically change our understanding of epidemiology [9,10]. However, despite *C. perfringens'* worldwide distribution, limited information is available on the genome of this pathogen, particularly in Saudi Arabia. Genes associated with antimicrobial resistance in the genome of the clinical isolate *C. perfringens* IRMC2505A could account for multidrug resistance. We used the genome analysis of this clinical isolate to test this hypothesis and investigate antimicrobial resistance-related genes involved in different resistance mechanisms. The genome of *C. perfringens* IRMC2505A was completely sequenced and analyzed using bioinformatics.

## 2. Results

The assembled genome of IRMC2505A had 108 contigs with a total length of 4,322,225 bp and an average G + C content of 28.28% (Table S1). This genome belongs to the Bacteria superkingdom, and its taxonomy is cellular organisms > Bacteria > Terrabacteria group > Firmicutes > Clostridia > Eubacterial > Clostridiaceae > *Clostridium* > *Clostridium perfringens*. The *Clostridium perfringens* genome has 4159 protein-coding sequences (CDS), 80 transfer RNA (tRNA) genes, and 7 ribosomal RNA (rRNA) genes (Table S1). Figure 1 shows a circular graphic of the distribution of *Clostridium perfringens* IRMC2505A genome annotations. It also presents an overview of the genome's specific biological processes and structural complexes. The IRMC2505A genome is close to *Clostridium perfringens* based on type (Tables 1 and 2), functional categorization (Figure 1 and Figure S1), and the number of specialty genes. We also performed a phylogenetic tree and genome comparison between *Clostridium perfringens* IRMC2505A and *Clostridium perfringens* ATCC 13124 195103.10 (Figure 2 and Figure S2). However, the genome comparison between IRMC2505A and ATCC 13124 195103.10 revealed dissimilar arrangements (Figure 2B). Protein analysis revealed the presence of 2748 proteins with functional assignments and 1411 hypothetical proteins (Table S1). The proteins with functional assignments included 870 proteins with EC numbers and 692 proteins with GO assignments. The *Clostridium perfringens* IRMC2505A genome has 3608 genus-specific protein families (PLFams) and 3702 cross-genus protein families (PGFams).

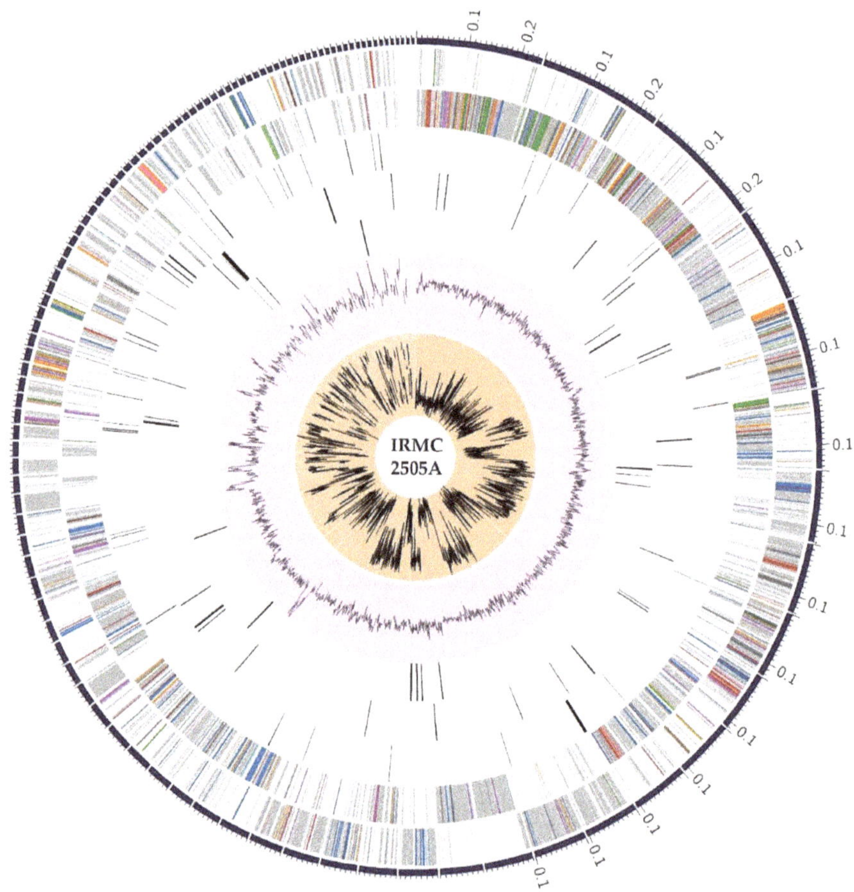


**Figure 1.** A circular graphic of the distribution of genome annotations in IRMC2505A. From the outer to inner rings, this includes the contigs, CDS on the forward strand, CDS on the reverse strand, RNA genes, CDS with homology to known antimicrobial resistance genes, CDS with homology to known virulence factors, GC content, and GC skew. The colors of the CDS on the forward and reverse strands indicate the subsystem these genes belong to (see subsystems below).

**Table 1.** Mechanisms of antimicrobial-resistant genes identified in the *Clostridium perfringens* IRMC2505A genome.

| S. No. | Antimicrobial Resistant Mechanism | Genes |
|---|---|---|
| 1 | Antibiotic inactivation enzyme | *NimB* |
| 2 | Antibiotic target in susceptible species | *Alr, Ddl, dxr, EF-G, EF-Tu, folA, Dfr, folP, gyrA, gyrB, Iso-tRNA, kasA, MurA, rho, rpoB, rpoC, S10p, S12p* |
| 3 | Antibiotic target protection protein | *TetB(P)* |
| 4 | Antibiotic target replacement protein | *FabV* |
| 5 | Efflux pump conferring antibiotic resistance | *TetA(P)* |
| 6 | Gene conferring resistance via absence | *GidB* |
| 7 | Protein altering cell wall charge conferring antibiotic resistance | *GdpD, MprF, PgsA* |
| 8 | Protein involved in antibiotic sequestration | *FabK-like* |

**Table 2.** List of antimicrobial resistance genes identified in the *Clostridium perfringens* IRMC2505A genome.

| | Source | Gene | Product/Function | Classification | Subject Coverage | Query Coverage | Identity | E-Value |
|---|---|---|---|---|---|---|---|---|
| 1 | NDARO/CARD | *tetB(P)* | Tetracycline resistance, ribosomal protection type ≥ TetB(P) | antibiotic target protection protein, tetracycline-resistant gene | 13 | 91 | 95 | $5 \times 10^{-40}$ |
| 2 | CARD/NDARO | *tetB(P)* | Tetracycline resistance, ribosomal protection type ≥ TetB(P) | antibiotic target protection protein, tetracycline-resistant gene | 48 | 99 | 98 | $1 \times 10^{-180}$ |
| 3 | CARD/PATRIC | *EF-Tu* | Translation elongation factor Tu | antibiotic resistant gene variant or mutant, elfamycin-resistant gene/antibiotic target in susceptible species | 6 | 96 | 83 | $7 \times 10^{-3}$ |
| 4 | NDARO/CARD/PATRIC | *tetA(P)* | Tetracycline resistance, MFS efflux pump ≥ TetA(P) | efflux pump conferring antibiotic resistance | 100 | 100 | 98 | $1 \times 10^{-234}$ |
| 5 | CARD | *mprF* | putative membrane protein | antibiotic target modifying enzyme, peptide antibiotic resistance gene | 100 | 100 | 95 | 0.0 |

CARD: Comprehensive Antibiotic Resistance Database; NDARO: National Database of Antibiotic Resistant Organisms; PATRIC: Pathosystems Resource Integration Center. Complete list of antimicrobial resistance genes identified in the *Clostridium perfringens* IRMC2505A genome with classifications is listed in Supplementary Tables S2 and S3.

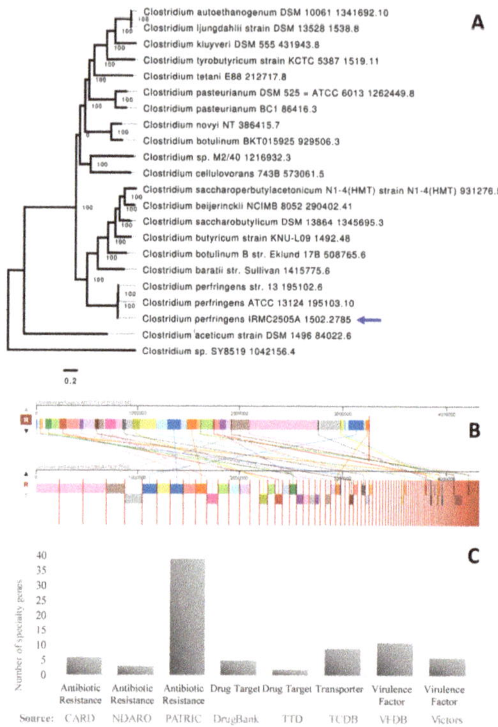

**Figure 2.** (**A**): Phylogenetic tree of the IRMC2505A genome. The blue arrow indicates the IRMC2505A genome. Tree Analysis Statistics: Requested genomes—22; Genomes with data—22; Max allowed deletions—0; Max allowed duplications—0; Single-copy genes requested—100; Single-copy genes found—100; Num protein alignments—100; Alignment program—mafft; Protein alignment time—189.9 s; Num aligned amino acids—46380; Num CDS alignments—100; Num aligned nucleotides—139140; Best protein model found by RAxML—LG; Branch support method—RAxML Fast Bootstrapping; RAxML likelihood—2571915.6763; RAxML version—8.2.11; RAxML time—3978.6 s. A list of genes used for tree analysis statistics is listed in Supplementary Table S4. (**B**): Genome comparison represents homologous regions between *Clostridium perfringens* IRMC2505A and ATCC 13124 195103.10. Alignable sequence blocks are common between the present study's clinical isolate (lower) and the reference (upper) *C. perfringens* ATCC 13124 195103.10. Homologous locally collinear blocks between IRMC2505A and ATCC 13124 195103.10 are identified by the same color and connected with a line. (**C**): Number of specialty genes and the specific source database where homology was found.

*Antimicrobial Resistance Genes*

The CARD (comprehensive antibiotic resistance database), NDARO (national database of antibiotic-resistant organisms), and PATRIC (Pathosystems Resource Integration Center) databases revealed antimicrobial resistance genes (Tables 1 and 2; Supplementary Tables S2 and S3). Nineteen (*Alr*, *Ddl*, *dxr*, *EF-G*, *EF-Tu*, *folA*, *Dfr*, *folP*, *gyrA*, *gyrB*, *Iso-tRNA*, *kasA*, *MurA*, *rho*, *rpoB*, *rpoC*, *S10p*, and *S12p*) antibiotic targets in susceptible species were identified in the genome of *Clostridium perfringens* IRMC2505A using k-mer-based detection methods for AMR genes (Table 1). Three genes (*GdpD*, *MprF*, and *PgsA*) involved in protein-altering cell wall charge conferring antibiotic resistance were also identified in the IRMC2505A genome. Only one gene in each category was associated with the resistant mechanism of the isolate *Clostridium perfringens:* antibiotic inactivation enzyme (*NimB*), antibiotic target protection protein (*TetB(P)*), antibiotic target replacement protein (*fabV*), efflux pump conferring antibiotic resistance (*TetA(P)*), gene conferring resistance via absence (*gidB*), and protein involved in antibiotic sequestration (*FabK-like*). The putative membrane protein (*mprF*) is similar to the genome of the comprehensive antibiotic resistance database (E-value = 0.0). Tetracycline-resistant MFS efflux pump (*TetA(P)*) is similar to the genome reported in the NDARO, CARD, and PATRIC databases (E-value = $1 \times 10^{-234}$) (Figure 3). We compared regional views of the putative membrane protein, tetracycline-resistant MFS efflux pump, and tetracycline-resistant ribosomal protection type genes in the IRMC2505A genome using the PATRIC cross-genus families (PGfams) method with reference genomes confirming similar subject coverage, query coverage, and identity (Table 2 and Figure 3). Seventeen virulence factors were identified in the *Clostridium perfringens* IRMC2505A genome (Table 3). More than 50% of the virulence factors are exoenzymes. Using the CARD and VFDB databases, metagenomic read mapping against the template genome revealed significant ($p$-value = $1 \times 10^{-26}$) genes with aligned reads against antibiotic-resistant genes or virulence factors, including phospholipase C, perfringolysin O, collagenase, hyaluronidase, alpha-clostripain, exo-alpha-sialidase, and sialidase activity (Table 4).

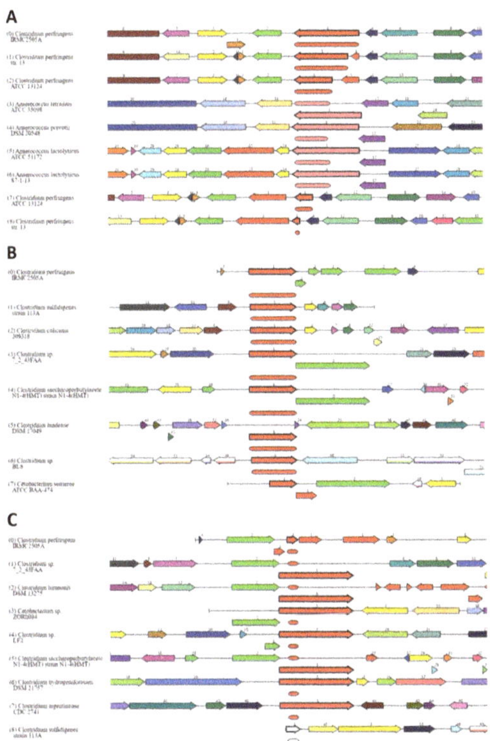

**Figure 3.** Regional view comparison of resistant genes identified in the IRMC2505A genome using the PATRIC cross-genus families (PGfams) method with reference genomes in the 10,000 bp region around the selected gene. Label 1 indicates the gene of interest. The forward and reverse orientations of genes in regions around the selected gene are indicated by arrows. Genes displaying identical colors indicate similar homology. (**A**): putative membrane protein (Classification: antibiotic target modifying enzyme, peptide antibiotic resistance gene), E-value = 0. (**B**): tetracycline-resistant MFS efflux pump ≥ TetA(P) (classification: efflux pump conferring antibiotic resistance), E-value: $1 \times 10^{-234}$. (**C**): Tetracycline-resistant ribosomal protection type ≥ TetB(P) (classification: antibiotic target protection protein, tetracycline-resistant gene), E-value: $5 \times 10^{-40}$.

**Table 3.** List of virulence factors identified in the *Clostridium perfringens* IRMC2505A genome.

| Source | Source ID | Source Organism | Gene | Product | Classification | Subject Coverage | Query Coverage | Identity | E-Value |
|---|---|---|---|---|---|---|---|---|---|
| Victors | 15676067 | *Neisseria meningitidis* MC58 | *tufA* | Translation elongation factor Tu | | 5 | 92 | 86 | $3 \times 10^{-3}$ |
| VFDB | VFG002285 | *Clostridium perfringens* ATCC 13124 | *nanH* | Sialidase (EC 3.2.1.18) | Exoenzyme, Carbohydrate-active enzyme, Sialidase | 100 | 100 | 98 | $1 \times 10^{-226}$ |
| Victors | 110801372 | *Clostridium perfringens* ATCC 13124 | *virS* | Hypothetical protein | | 99 | 100 | 95 | $1 \times 10^{-243}$ |
| Victors | 18309145 | *Clostridium perfringens* str. 13 | *pfoA* | Thiol-activated cytolysin | | 92 | 88 | 87 | $1 \times 10^{-243}$ |
| Victors | 15676067 | *Neisseria meningitidis* MC58 | *tufA* | Translation elongation factor Tu | | 5 | 92 | 86 | $3 \times 10^{-3}$ |
| VFDB | VFG002278 | *Clostridium perfringens* str. 13 | *nagI* | Protein O-GlcNAcase (EC 3.2.1.169) | Exoenzyme, Carbohydrate-active enzyme, Hyaluronidase | 100 | 100 | 96 | 0.0 |
| Victors | 29376182 | *Enterococcus faecalis* V583 | *EF1623* | Ethanolamine utilization protein similar to PduA/PduJ | | 91 | 91 | 92 | $4 \times 10^{-38}$ |
| VFDB | VFG002274 | *Clostridium perfringens* str. 13 | *plc* | Broad-substrate range phospholipase C (EC 3.1.4.3) | Toxin, Zinc-metallo-phospholipase C | 100 | 100 | 99 | $1 \times 10^{-242}$ |
| VFDB | VFG002282 | *Clostridium perfringens* str. 13 | *cloSI* | Hypothetical protein | Exoenzyme, Cysteine endopeptidase | 99 | 100 | 97 | $1 \times 10^{-306}$ |
| VFDB | VFG002279 | *Clostridium perfringens* str. 13 | *nagJ* | Protein O-GlcNAcase (EC 3.2.1.169) | Exoenzyme, Carbohydrate-active enzyme, Hyaluronidase | 100 | 100 | 98 | 0.0 |
| Victors | 18309018 | *Clostridium perfringens* str. 13 | *plc* | Broad-substrate range phospholipase C (EC 3.1.4.3) | | 100 | 100 | 99 | $1 \times 10^{-242}$ |
| VFDB | VFG002281 | *Clostridium perfringens* str. 13 | *nagL* | Hypothetical protein | Exoenzyme, Carbohydrate-active enzyme, Hyaluronidase | 100 | 100 | 95 | 0.0 |
| VFDB | VFG002283 | *Clostridium perfringens* str. 13 | *nanI* | Sialidase (EC 3.2.1.18) | Exoenzyme, Carbohydrate-active enzyme, Sialidase | 100 | 100 | 99 | 0.0 |
| VFDB | VFG002276 | *Clostridium perfringens* str. 13 | *colA* | Microbial collagenase (EC 3.4.24.3) | Exoenzyme, Collagenase | 100 | 100 | 98 | 0.0 |
| VFDB | VFG002284 | *Clostridium perfringens* str. 13 | *nanJ* | Sialidase (EC 3.2.1.18) | Exoenzyme, Carbohydrate-active enzyme, Sialidase | 100 | 100 | 95 | 0.0 |
| VFDB | VFG002275 | *Clostridium perfringens* str. 13 | *pfoA* | Thiol-activated cytolysin | Toxin, Membrane-damaging, Pore-forming, Channel-forming involving beta-sheet-containing toxin (beta-barrel), Cholesterol-dependent cytolysin | 92 | 88 | 87 | $1 \times 10^{-243}$ |
| VFDB | VFG002277 | *Clostridium perfringens* str. 13 | *nagH* | Hyalurono-glucosaminidase (EC 3.2.1.35) | Exoenzyme, Carbohydrate-active enzyme, Hyaluronidase | 100 | 100 | 97 | 0.0 |

**Table 4.** Metagenomic read mapping of *Clostridium perfringens* IRMC2505A against the templates using the CARD and VFDB databases.

| Template | Gene Function | Genome | Score | Expected | Template Length | Template Identity | Template Coverage | Query Identity | Query Coverage | Depth | q-Value | *p*-Value |
|---|---|---|---|---|---|---|---|---|---|---|---|---|
| CARD∣ABG86067.1 | *Clostridium perfringens* mprF | $ | 1395 | 1 | 1710 | 93.8 | 100 | 93.8 | 100 | 1 | 1390.34 | $1.00 \times 10^{-26}$ |
| CARD∣AAA20116.1 | tetA(P) | # | 1218 | 1 | 1263 | 98.73 | 100 | 98.73 | 100 | 1 | 1214.3 | $1.00 \times 10^{-26}$ |
| VFDB∣VFG002274 | phospholipase C | * | 1164 | 7 | 1197 | 99 | 100 | 99 | 100 | 1 | 1140.8 | $1.00 \times 10^{-26}$ |
| VFDB∣VFG002275 | perfringolysin O | * | 915 | 9 | 1503 | 86.56 | 98.27 | 88.08 | 101.76 | 0.98 | 885.75 | $1.00 \times 10^{-26}$ |
| VFDB∣VFG002276 | collagenase | * | 3177 | 19 | 3315 | 98.58 | 100 | 98.58 | 100 | 1 | 3117.65 | $1.00 \times 10^{-26}$ |
| VFDB∣VFG002277 | hyaluronidase | * | 4467 | 27 | 4887 | 97.11 | 100 | 97.11 | 100 | 1 | 4384.16 | $1.00 \times 10^{-26}$ |
| VFDB∣VFG002278 | hyaluronidase | * | 3537 | 23 | 3894 | 96.92 | 100 | 96.92 | 100 | 1 | 3468.34 | $1.00 \times 10^{-26}$ |
| VFDB∣VFG002279 | hyaluronidase | * | 2769 | 18 | 3006 | 97.34 | 100 | 97.34 | 100 | 1 | 2714.3 | $1.00 \times 10^{-26}$ |
| VFDB∣VFG002281 | hyaluronidase | * | 2976 | 20 | 3384 | 95.95 | 100 | 95.95 | 100 | 1 | 2914.96 | $1.00 \times 10^{-26}$ |
| VFDB∣VFG002282 | alpha-clostripain | * | 1459 | 10 | 1575 | 97.65 | 100 | 97.65 | 100 | 1 | 1428.83 | $1.00 \times 10^{-26}$ |
| VFDB∣VFG002283 | exo-alpha-sialidase | * | 1995 | 13 | 2085 | 98.51 | 100 | 98.51 | 100 | 1 | 1955.87 | $1.00 \times 10^{-26}$ |
| VFDB∣VFG002284 | exo-alpha-sialidase | * | 3206 | 21 | 3522 | 97.02 | 100 | 97.02 | 100 | 1 | 3143.05 | $1.00 \times 10^{-26}$ |
| VFDB∣VFG002285 | sialidase | ‡ | 1026 | 7 | 1149 | 96.34 | 100 | 96.34 | 100 | 1 | 1003.63 | $1.00 \times 10^{-26}$ |

Template: reference gene sequence; Genome: Genome that contains template gene; $: *Clostridium perfringens* SM101; #: *Clostridium perfringens*; *: *Clostridium perfringens* str. 13; ‡: *Clostridium perfringens* ATCC 13124. Score: global alignment score of the template; Expected: expected alignment score; Template length: Template gene length in nucleotides; Template Identity: Percentage (%) of identity between the query and template over the length of the matching query; template Coverage: Percentage (%) of the template that is covered by the query; Query Identity: Percentage (%) of identity between the query and template sequence over the length of the matching query sequence; Query Coverage: length of the matching query divided by the template length; Depth: Number of times the template is covered by the query; q-value: Quantile from McNemar's test to evaluate whether the current template is significant; *p*-value: *p*-value corresponding to the obtained q-value.

## 3. Discussion

Although many food-borne diseases are surveyed and controlled, and several enteric organisms are monitored through databases, *Clostridium perfingens* remains one of the only toxicogenic clostridial species identified by local investigations after an outbreak occurs. The bacterium is reported worldwide and has been listed among the leading species of foodborne and non-foodborne gastroenteritis [11]. Although a serious threat is imposed by the rise of multi-drug resistant (MDR)/toxigenic strains, little is known about the prevalence and genetic composition of this pathogen in the Middle East. To our knowledge, the few studies conducted in Saudi Arabia are limited to livestock and raw animal meats. None of them involve whole genome sequencing for species identification or studying antimicrobial resistance and toxic behavioral patterns in clostridial isolates from humans [12,13].

This genomic study of *C. perfingens* was isolated from a patient in Saudi Arabia's Eastern Province. It would be beneficial to start a surveillance program and elucidate changes in the pathogen's genome. This information is clinically significant since the antimicrobial resistance profile and genome characteristics of these isolates are still understudied in Saudi Arabia. Additionally, recent studies have identified hypervirulent clostridial spores in new environments and genomically linked them to new matrices other than hospital environments due to uncontrolled antimicrobial usage [14,15]. Our findings may also provide insight into the adaptation of these bacteria to new niches.

We believe that a better understanding of *C. perfingens* resistance mechanisms could decrease the misuse of broad-spectrum antibiotics and AMR genes' potential transfer to other pathogens.

### 3.1. Phylogenetic Analyses

Phylogenetic analysis confirmed the phylotype and found 100% similarity between our isolate sequences, *C. perfingens* ATCC 13124 195103.10, and *C. perfringens* str. 13 195102.6 species.

The last species, *C. perfringens* str. 13, was sequenced for the first time in 2002 by Shimizu et al. It is a natural isolate found in the soil and classified as a type A strain [16]. Genome annotation distribution in IRMC2505A showed that the average genome length, GC content, and rRNA copy number were 4,322,225 bp, 28.28%, and 7, respectively. This anaerobic organism had several genes belonging to metabolism (523), protein processing (238), virulence (100), cellular processes (179), energy (135), and DNA and RNA processing (107,61). A recent study claimed that *C. perfringens* has the highest genome plasticity (by calculating the openness index (OI)) out of 51 other bacterial species. In other words, this bacterium has a highly open genome to acquire new adaptive genes for survival in hostile external environments. These properties are reflected by its low GC content (around 28%) and high 16S rRNA gene copy number [17].

### 3.2. Antimicrobial Resistance

Many research studies have indicated the emergence of multi-resistant strains of clostridial species and their significant risk to humans. In this study, the antimicrobial resistance profile of *C. perfingens* showed a multi-drug-resistant isolate involving nineteen resistance genes with different resistance mechanisms. Most of these genes were in the "antibiotic target in susceptible species" category. Many of these identical genes, with their antimicrobial resistance mechanisms, have been identified and described in different gram-negative bacteria where homologous combinations are found [18].

Predicting AMR genes based on NDARO/CARD/PATRIC databases showed tetracycline resistance (% query coverage and % identity > 97%) and defensin-like cationic antimicrobial peptides (CAMPs) (% query coverage = 100% and % identity = 95%). We detected two tetracycline-resistant genes, *tetA(P)* and *tetB(P)*, which are known to encode proteins that mediate active tetracycline efflux and ribosomal protection tetracycline-resistant mechanisms, respectively [19]. The anti-defensins gene (*mprF*) reported codes for multiple peptide resistance factor proteins, which are involved in CAMP resistance [20]. Moreover, elfamycin-resistant genes were also detected with high query coverage (96%) but displayed an identity of 83%. These proteins are known as inhibitors of the Elongation Factor-Thermo Unstable (EF-Tu) used in the cell to translate mRNA transcripts into proteins [21].

Antimicrobial agents are widely used in animals as growth promoters and can lead to the emergence of antimicrobial-resistant enteric microflora, including *C. perfringens* and its spores [22]. Hence, antibiotic-resistant *C. perfringens* should be investigated in many matrixes (animal, human, food, soil ... ) [23]. Motamedi et al. conducted the first systematic review in 2021 on *C. perfringens* as a causative agent of antibiotic-associated diarrhea (AAD) in hospitalized patients [24]. These studies were conducted mainly in Asia, Europe, and Africa and showed 14.9% *C. perfringens* among patients. However, insufficient data were found in these areas. Consequently, this finding stresses the urgent need for a surveillance program to control AAD and guarantee efficient treatment.

A similar situation was observed in Saudi Arabia, where a recent meta-analysis (2022) on the frequency of antibiotic-resistant *Clostridium* species found no evidence for the antimicrobial resistance patterns reported in KSA. Most of these limited studies used conventional methods (selective culture on agar media, E-tests, and disk diffusion assays) [25]. These reference methods only isolate and identify the studied bacteria and do not distinguish between different pathogens based on their virulence phenotypes. Limited research studies in KSA investigated the resistance and occurrence of *C. perfringens* in camel meat. They showed these bacteria's high resistance to tetracycline and ceftiofur drugs and, to a lesser extent, other antimicrobial agents (erythromycin, oxytetracycline, penicillin, metronidazole, clindamycin, and lincomycin) [12,26,27].

### 3.3. Toxigenicity

*C. perfringens* is also known for its toxigenicity. Five toxinotypes (A–G) were historically identified based on the toxin produced by the strain (alpha, beta, epsilon, and iota toxins). Afterward, two new toxinotypes (F and G) were added to the scheme, introducing enterotoxin CPE (but not CPB, ETX, or ITX) and NetB toxin, respectively [28]. Many virulence factors contribute to this pathogen's toxicity. *C. perfringens* is known to produce a wide range of degradative enzymes, including hydrolytic enzymes, membrane-damaging enzymes, pore-forming toxins, and intracellular toxins (proteases, hyaluronidase, collagenase, endoglycosidases, etc.). The virulence analysis of our bacterial isolate showed a variable array of virulence genes and a 97% similarity to the *C. perfringens* genome. High query coverage sequences (≥97%) were distributed as follows: Sequences correlated to phospholipase C (*plc*) encoding the alpha toxin, sialidase genes (*nanH, nanI*), alpha-clostripain genes (*cloSI*), hyaluronidase genes (*nagJ, nagH*), and collagenase genes (*colA*) encoding the κ-toxin. These genes are chromosomal and located specifically in the conserved region of the *C. perfringens* chromosome [29]. There has been no WGS study on *C. perfringens* in Saudi Arabia, including the Arabian Peninsula [30].

In conclusion, this WGS study of *C. perfringens* showed an MDR bacterium with several virulence factors. To our knowledge, our WGS study is one of the first studies conducted in Saudi Arabia and the Arabian Peninsula on *C. perfringens*, where national databases are still inexistent. To guarantee a complete understanding of *C. perfringens'* epidemiology, a full study of this pathogen's virulence factors, and antimicrobial resistance patterns is essential. Future studies will also be beneficial in developing an effective disease control strategy.

## 4. Materials and Methods

### *4.1. Isolation and DNA Extraction of* C. perfringens *IRMC2505A*

This project was submitted for ethical approval to the ethical committee of Imam Abdulrahman Bin Faisal University (IRB-2022-01-398). All procedures were performed in accordance with the 1964 Helsinki Declaration and its later amendments or comparable ethical standards.

*A Clostridium perfringens* strain was isolated using cycloserine, cefoxitin, and fructose agar medium from the fecal sample of a patient who was clinically suspected of gastrointestinal infection and had a recent history of antibiotic exposure and diarrhea. A positive sample, tested for both toxins A and B by GeneXpert, was cultured on cycloserine cefoxitin fructose agar selective medium (CCFA) (MOLEQULE-ON, Auckland, New Zealand) under anaerobic conditions at 37 °C for 24 h. The total DNA was extracted using the Gentra Puregene Yeast/Bact. Kit (Qiagen, Hilden, Germany). The purity, quality, and quantity of genomic DNA were measured using Nanodrop 2000 (Thermo Scientific, Waltham, MA, USA) as described in the manufacturer's instructions. The isolate was PCR amplified, and *16S rRNA* was sequenced to confirm the isolate.

### *4.2. Virulence Genes Identification Using Whole Genome Sequencing*

The whole genome of isolate IRMC2505A was sequenced using the Illumina platform. The paired reads were assembled and annotated using PATRIC (BV-BRC 3.28.5) [31] and the RAST tool kit (RASTtk 1.3.0) [32]. We calculated the number of contigs and average G + C content and identified the taxonomy of the genome. We also predicted the proteins and their functional assignments with the Enzyme Commission (EC) [33], Gene Ontology (GO) [34], pathways [35], protein family types [36], and subsystems of protein complexes [37]. Specialty genes in the genome of IRMC2505A were identified using specific source databases for known transporters [38], virulence factors [39,40], drug targets [41,42], and antibiotic-resistant genes [43]. We detected k-mer-based antimicrobial resistance (AMR) genes [31]. We phylogenetically analyzed the IRMC2505A genome using 100 genes from the National Center for Biotechnology Information (NCBI) reference and representative genomes by Mash/MinHash (Mash v2.3) and PGFams (PATRIC, BV-BRC 3.28.5) [36] aligned with MUSCLE [44], a matrix analysis using RaxML (version 8) with fast bootstrapping [45,46]. Metagenomic read mapping against the selected template was analyzed using the CARD (2020) and VFDB (2019) databases through k-mer alignment [47].

**Supplementary Materials:** The following supporting information can be downloaded at: https://www.mdpi.com/article/10.3390/toxins15060359/s1, Figure S1: A: A circular graphical display of the distribution of the genome annotations in IRMC2505A. B: An overview of the subsystems of genes for the genome IRMC2505A; Figure S2: Proteome comparison. List of tracks, from outside to inside: *Clostridium perfringens* ATCC 13124 (195103.10) and *Clostridium perfringens* IRMC2505A (1502.2785); Table S1: Assembly Details and annotated features of *Clostridium perfringens* IRMC2505A; Table S2: List of antimicrobial resistance genes identified in the genome of *Clostridium perfringens* IRMC2505A; Table S3: Functions and classifications of list of antimicrobial resistance genes in the genome of *C. perfringens* IRMC2505A; Table S4: Gene families are ranked by alignment score for phylogenetic tree of IRMC2505A genome.

**Author Contributions:** R.A., D.M.A. and J.F.B. conceived and designed the research; R.A., D.M.A., N.B.A., S.A. and J.F.B. performed and analysed the experiments; S.A. and J.F.B. performed the whole-genome analysis; R.A., D.M.A. and M.F. performed clinical analysis; R.A., D.M.A., M.F., N.B.A., S.A. and J.F.B. wrote, reviewed and approved the final manuscript. All authors have read and agreed to the published version of the manuscript.

**Funding:** This research received no external funding.

**Institutional Review Board Statement:** The study was conducted in accordance with the Declaration of Helsinki, and approved by the Institutional Review Board of Imam Abdulrahman Bin Faisal University (IRB-2022-01-398).

**Informed Consent Statement:** Not applicable.

**Data Availability Statement:** All data will be available on reasonable request from the corresponding author.

**Conflicts of Interest:** The authors declare no conflict of interest.

## References


1. Ueda, K.; Kawahara, K.; Kimoto, N.; Yamaguchi, Y.; Yamada, K.; Oki, H.; Yoshida, T.; Matsuda, S.; Matsumoto, Y.; Motooka, D.; et al. Analysis of the complete genome sequences of *Clostridium perfringens* strains harbouring the binary enterotoxin BEC gene and comparative genomics of pCP13-like family plasmids. *BMC Genom.* **2022**, *23*, 226. [CrossRef] [PubMed]
2. Chen, J.; Ma, M.; Uzal, F.A.; McClane, B.A. Host cell-induced signaling causes *Clostridium perfringens* to upregulate production of toxins important for intestinal infections. *Gut. Microbes* **2013**, *5*, 96–107. [CrossRef] [PubMed]
3. Kiu, R.; Caim, S.; Alexander, S.; Pachori, P.; Hall, L.J. Probing genomic aspects of the multi-host pathogen *Clostridium perfringens* reveals significant pangenome diversity, and a diverse array of virulence factors. *Front. Microbiol.* **2017**, *8*, 2485. [CrossRef] [PubMed]
4. Hassan, K.A.; Elbourne, L.D.H.; Tetu, S.G.; Melville, S.B.; Rood, J.I.; Paulsen, I.T. Genomic analyses of *Clostridium perfringens* isolates from five toxinotypes. *Res. Microbiol.* **2015**, *166*, 255–263. [CrossRef]
5. Sakurai, J.; Nagahama, M.; Oda, M. *Clostridium perfringens* Alpha-Toxin: Characterization and Mode of Action. *J. Biochem.* **2004**, *136*, 569–574. [CrossRef]
6. Kubota, H.; Sakai, T.; Gawad, A.; Makino, H.; Akiyama, T.; Ishikawa, E.; Oishi, K. Development of TaqMan-Based Quantitative PCR for Sensitive and Selective Detection of Toxigenic Clostridium difficile in Human Stools. *PLoS ONE* **2014**, *9*, e111684. [CrossRef]
7. Miyamoto, K.; Li, J.; McClane, B.A. Enterotoxigenic *Clostridium perfringens*: Detection and identification. *Microbes Environ.* **2012**, *27*, 343–349. [CrossRef]
8. Bennett, S.D.; Walsh, K.A.; Gould, L.H. Foodborne Disease Outbreaks Caused by *Bacillus cereus*, *Clostridium perfringens*, and *Staphylococcus aureus*—United States, 1998–2008. *Clin. Infect. Dis.* **2013**, *57*, 425–433. [CrossRef]
9. Martin, J.S.H.; Monaghan, T.M.; Wilcox, M.H. Clostridium difficile infection: Epidemiology, diagnosis and understanding transmission. *Nat. Rev. Gastroenterol. Hepatol.* **2016**, *13*, 206–216. [CrossRef]
10. Eyre, D.W.; Walker, A.S. Clostridium difficile surveillance: Harnessing new technologies to control transmission. *Expert Rev. Anti-Infect. Ther.* **2014**, *11*, 1193–1205. [CrossRef]
11. *C. perfringens* | CDC. Available online: https://www.cdc.gov/foodsafety/diseases/clostridium-perfringens.html (accessed on 5 February 2023).
12. Omer, S.A.; Al-Olayan, E.M.; Babiker, S.E.H.; Aljulaifi, M.Z.; Alagaili, A.N.; Mohammed, O.B. Genotyping of *Clostridium perfringens* Isolates from Domestic Livestock in Saudi Arabia. *BioMed Res. Int.* **2020**, *2020*, 9035341. [CrossRef] [PubMed]
13. Bendary, M.M.; Abd El-Hamid, M.I.; El-Tarabili, R.M.; Hefny, A.A.; Algendy, R.M.; Elzohairy, N.A.; Ghoneim, M.M.; Al-Sanea, M.M.; Nahari, M.H.; Moustafa, W.H. *Clostridium perfringens* Associated with Foodborne Infections of Animal Origins: Insights into Prevalence, Antimicrobial Resistance, Toxin Genes Profiles, and Toxinotypes. *Biology* **2022**, *11*, 551. [CrossRef] [PubMed]

14. Lim, S.C.; Collins, D.A.; Imwattana, K.; Knight, D.R.; Perumalsamy, S.; Hain-Saunders, N.M.R.; Putsathit, P.; Speers, D.; Riley, T.V. Whole-genome sequencing links Clostridium (Clostridioides) difficile in a single hospital to diverse environmental sources in the community. *J. Appl. Microbiol.* **2022**, *133*, 1156–1168. [CrossRef] [PubMed]
15. Mitchell, M.; Nguyen, S.V.; MacOri, G.; Bolton, D.; McMullan, G.; Drudy, D.; Fanning, S. *Clostridioides difficile* as a Potential Pathogen of Importance to One Health: A Review. *Foodborne Pathog. Dis.* **2022**, *19*, 806–816. [CrossRef]
16. Shimizu, T.; Ohtani, K.; Hirakawa, H.; Ohshima, K.; Yamashita, A.; Shiba, T.; Ogasawara, N.; Hattori, M.; Kuhara, S.; Hayashi, H. Complete genome sequence of *Clostridium perfringens*, an anaerobic flesh-eater. *Proc. Natl. Acad. Sci. USA* **2002**, *99*, 996–1001. [CrossRef]
17. Feng, Y.; Fan, X.; Zhu, L.; Yang, X.; Liu, Y.; Gao, S.; Jin, X.; Liu, D.; Ding, J.; Guo, Y.; et al. Phylogenetic and genomic analysis reveals high genomic openness and genetic diversity of *Clostridium perfringens*. *Microb. Genom.* **2020**, *6*, e000441. [CrossRef]
18. Arefiev, V.; Kovalenko, G.; Frant, M.; Chumachenko, T.; Polyvianna, Y.; Pivnenko, S.; Bolotin, V.; Mayboroda, O.; Solodiankin, O.; Tarasov, O.; et al. Complete Genome Sequence of *Salmonella enterica* subsp. *enterica* Serovar Kottbus Strain Kharkiv, Isolated from a Commercial Pork Production Facility in Ukraine. *Microbiol. Resour. Announc.* **2020**, *9*, e01171-20. [CrossRef]
19. Sloan, J.; McMurry, L.M.; Lyras, D.; Levy, S.B.; Rood, J.I. The *Clostridium perfringens* Tet P determinant comprises two overlapping genes: TetA(P), which mediates active tetracycline efflux, and tetB(P), which is related to the ribosomal protection family of tetracycline-resistance determinants. *Mol. Microbiol.* **1994**, *11*, 403–415. [CrossRef]
20. Ernst, C.M.; Staubitz, P.; Mishra, N.N.; Yang, S.-J.; Hornig, G. The Bacterial Defensin Resistance Protein MprF Consists of Separable Domains for Lipid Lysinylation and Antimicrobial Peptide Repulsion. *PLoS Pathog.* **2009**, *5*, 1000660. [CrossRef]
21. Prezioso, S.M.; Brown, N.E.; Goldberg, J.B. Elfamycins: Inhibitors of elongation factor-Tu. *Mol. Microbiol.* **2017**, *106*, 22–34. [CrossRef]
22. Odeyemi, O.A.; Sani, N.A. Antibiotic resistance and burden of foodborne diseases in developing countries. *Futur. Sci. OA* **2016**, *2*, FSO139. [CrossRef] [PubMed]
23. Park, M.; Rooney, A.P.; Hecht, D.W.; Li, J.; McClane, B.A.; Nayak, R.; Paine, D.D.; Rafii, F. Phenotypic and genotypic characterization of tetracycline and minocycline resistance in *Clostridium perfringens*. *Arch. Microbiol.* **2010**, *192*, 803–810. [CrossRef] [PubMed]
24. Motamedi, H.; Fathollahi, M.; Abiri, R.; Kadivarian, S.; Rostamian, M.; Alvandi, A. A worldwide systematic review and metaanalysis of bacteria related to antibioticassociated diarrhea in hospitalized patients. *PLoS ONE* **2021**, *16*, e0260667. [CrossRef]
25. Banawas, S.S. Systematic Review and Meta-Analysis on the Frequency of Antibiotic-Resistant Clostridium Species in Saudi Arabia. *Antibiotics* **2022**, *11*, 1165. [CrossRef]
26. Fayez, M.; Elsohaby, I.; Al-Marri, T.; Zidan, K.; Aldoweriej, A.; El-Sergany, E.; Elmoslemany, A. Genotyping and antimicrobial susceptibility of *Clostridium perfringens* isolated from dromedary camels, pastures and herders. *Comp. Immunol. Microbiol. Infect. Dis.* **2020**, *70*, 101460. [CrossRef] [PubMed]
27. Fayez, M.; El-Ghareeb, W.R.; Elmoslemany, A.; Alsunaini, S.J.; Alkafafy, M.; Alzahrani, O.M.; Mahmoud, S.F.; Elsohaby, I. Genotyping and Antimicrobial Susceptibility of *Clostridium perfringens* and *Clostridioides difficile* in Camel Minced Meat. *Pathogens* **2021**, *10*, 1640. [CrossRef] [PubMed]
28. Rood, J.I.; Adams, V.; Lacey, J.; Lyras, D.; McClane, B.A.; Melville, S.B.; Moore, R.J.; Popoff, M.R.; Sarker, M.R.; Songer, J.G.; et al. Expansion of the *Clostridium perfringens* toxin-based typing scheme. *Anaerobe* **2018**, *53*, 5–10. [CrossRef]
29. Abdelrahim, A.M.; Radomski, N.; Delannoy, S.; Djellal, S.; Le Négrate, M.; Hadjab, K.; Fach, P.; Hennekinne, J.A.; Mistou, M.Y.; Firmesse, O. Large-Scale genomic analyses and toxinotyping of *Clostridium perfringens* implicated in foodborne outbreaks in France. *Front. Microbiol.* **2019**, *10*, 777. [CrossRef]
30. Francis Borgio, J.; Rasdan, A.S.; Sonbol, B.; Alhamid, G.; Almandil, N.B.; Azeez, S.A. Emerging status of multidrug-resistant bacteria and fungi in the arabian peninsula. *Biology* **2021**, *10*, 1144. [CrossRef]
31. Wattam, A.R.; Davis, J.J.; Assaf, R.; Boisvert, S.; Brettin, T.; Bun, C.; Conrad, N.; Dietrich, E.M.; Disz, T.; Gabbard, J.L.; et al. Improvements to PATRIC, the all-bacterial Bioinformatics Database and Analysis Resource Center. *Nucleic Acids Res.* **2017**, *45*, D535–D542. [CrossRef]
32. Brettin, T.; Davis, J.J.; Disz, T.; Edwards, R.A.; Gerdes, S.; Olsen, G.J.; Olson, R.; Overbeek, R.; Parrello, B.; Pusch, G.D.; et al. RASTtk: A modular and extensible implementation of the RAST algorithm for building custom annotation pipelines and annotating batches of genomes. *Sci. Rep.* **2015**, *5*, 8365. [CrossRef] [PubMed]
33. Schomburg, I.; Chang, A.; Ebeling, C.; Gremse, M.; Heldt, C.; Huhn, G.; Schomburg, D. BRENDA, the enzyme database: Updates and major new developments. *Nucleic Acids Res.* **2004**, *32*, D431–D433. [CrossRef] [PubMed]
34. Gene Ontology Consortium, T.; Ashburner, M.; Ball, C.A.; Blake, J.A.; Botstein, D.; Butler, H.; Michael Cherry, J.; Davis, A.P.; Dolinski, K.; Dwight, S.S.; et al. Gene Ontology: Tool for the unification of biology NIH Public Access Author Manuscript. *Nat. Genet.* **2000**, *25*, 25–29. [CrossRef]
35. Kanehisa, M.; Sato, Y.; Kawashima, M.; Furumichi, M.; Tanabe, M. KEGG as a reference resource for gene and protein annotation. *Nucleic Acids Res.* **2016**, *44*, D457–D462. [CrossRef] [PubMed]
36. Davis, J.J.; Gerdes, S.; Olsen, G.J.; Olson, R.; Pusch, G.D.; Shukla, M.; Vonstein, V.; Wattam, A.R.; Yoo, H. PATtyFams: Protein families for the microbial genomes in the PATRIC database. *Front. Microbiol.* **2016**, *7*, 118. [CrossRef]

37. Overbeek, R.; Begley, T.; Butler, R.M.; Choudhuri, J.V.; Chuang, H.Y.; Cohoon, M.; de Crécy-Lagard, V.; Diaz, N.; Disz, T.; Edwards, R.; et al. The Subsystems Approach to Genome Annotation and its Use in the Project to Annotate 1000 Genomes. *Nucleic Acids Res.* **2005**, *33*, 5691–5702. [CrossRef]
38. Saier, M.H.; Reddy, V.S.; Tsu, B.V.; Ahmed, M.S.; Li, C.; Moreno-Hagelsieb, G. The Transporter Classification Database (TCDB): Recent advances. *Database Issue Publ.* **2016**, *44*, D372–D379. [CrossRef]
39. Mao, C.; Abraham, D.; Wattam, A.R.; Wilson, M.J.C.; Shukla, M.; Yoo, H.S.; Sobral, B.W. Curation, integration and visualization of bacterial virulence factors in PATRIC. *Bioinformatics* **2015**, *31*, 252–258. [CrossRef]
40. Chen, L.; Zheng, D.; Liu, B.; Yang, J.; Jin, Q. VFDB 2016: Hierarchical and refined dataset for big data analysis-10 years on. *Nucleic Acids Res.* **2016**, *44*, D694–D697. [CrossRef]
41. Zhu, F.; Han, B.; Kumar, P.; Liu, X.; Ma, X.; Wei, X.; Huang, L.; Guo, Y.; Han, L.; Zheng, C.; et al. Update of TTD: Therapeutic Target Database. *Nucleic Acids Res.* **2010**, *38*, D787–D791. [CrossRef]
42. Law, V.; Knox, C.; Djoumbou, Y.; Jewison, T.; Guo, A.C.; Liu, Y.; Maclejewski, A.; Arndt, D.; Wilson, M.; Neveu, V.; et al. DrugBank 4.0: Shedding new light on drug metabolism. *Nucleic Acids Res.* **2014**, *42*, D1091–D1097. [CrossRef] [PubMed]
43. McArthur, A.G.; Waglechner, N.; Nizam, F.; Yan, A.; Azad, M.A.; Baylay, A.J.; Bhullar, K.; Canova, M.J.; De Pascale, G.; Ejim, L.; et al. The comprehensive antibiotic resistance database. *Antimicrob. Agents Chemother.* **2013**, *57*, 3348–3357. [CrossRef] [PubMed]
44. Edgar, R.C. MUSCLE: Multiple sequence alignment with high accuracy and high throughput. *Nucleic Acids Res.* **2004**, *32*, 1792–1797. [CrossRef] [PubMed]
45. Stamatakis, A.; Hoover, P.; Rougemont, J. A Rapid Bootstrap Algorithm for the RAxML Web Servers. *Syst. Biol.* **2008**, *57*, 758–771. [CrossRef] [PubMed]
46. Stamatakis, A. RAxML version 8: A tool for phylogenetic analysis and post-analysis of large phylogenies. *Bioinformatics* **2014**, *30*, 1312–1313. [CrossRef]
47. Jia, B.; Raphenya, A.R.; Alcock, B.; Waglechner, N.; Guo, P.; Tsang, K.K.; Lago, B.A.; Dave, B.M.; Pereira, S.; Sharma, A.N. CARD 2017: Expansion and model-centric curation of the comprehensive antibiotic resistance database. *Nucleic Acids Res.* **2016**, *45*, D566–D573. [CrossRef]

toxins MDPI

*Article*

# Domperidone Protects Cells from Intoxication with *Clostridioides difficile* Toxins by Inhibiting Hsp70-Assisted Membrane Translocation

**Maria Braune-Yan, Jinfang Jia, Mary Wahba, Johannes Schmid, Panagiotis Papatheodorou, Holger Barth and Katharina Ernst ***

Institute of Experimental and Clinical Pharmacology, Toxicology and Pharmacology of Natural Products, Ulm University Medical Center, 89081 Ulm, Germany
* Correspondence: katharina.ernst@uni-ulm.de

**Abstract:** *Clostridioides difficile* infections cause severe symptoms ranging from diarrhea to pseudomembranous colitis due to the secretion of AB-toxins, TcdA and TcdB. Both toxins are taken up into cells through receptor-mediated endocytosis, autoproteolytic processing and translocation of their enzyme domains from acidified endosomes into the cytosol. The enzyme domains glucosylate small GTPases such as Rac1, thereby inhibiting processes such as actin cytoskeleton regulation. Here, we demonstrate that specific pharmacological inhibition of Hsp70 activity protected cells from TcdB intoxication. In particular, the established inhibitor VER-155008 and the antiemetic drug domperidone, which was found to be an Hsp70 inhibitor, reduced the number of cells with TcdB-induced intoxication morphology in HeLa, Vero and intestinal CaCo-2 cells. These drugs also decreased the intracellular glucosylation of Rac1 by TcdB. Domperidone did not inhibit TcdB binding to cells or enzymatic activity but did prevent membrane translocation of TcdB's glucosyltransferase domain into the cytosol. Domperidone also protected cells from intoxication with TcdA as well as CDT toxin produced by hypervirulent strains of *Clostridioides difficile*. Our results reveal Hsp70 requirement as a new aspect of the cellular uptake mechanism of TcdB and identified Hsp70 as a novel drug target for potential therapeutic strategies required to combat severe *Clostridioides difficile* infections.

**Keywords:** toxin uptake; chaperones; Hsp70; domperidone; *Clostridioides difficile* toxins; toxin inhibitors

**Key Contribution:** *Clostridioides difficile* infections cause severe symptoms due to the secretion of AB-toxins, TcdA and TcdB, which glucosylate small GTPases such as Rac1 and inhibit processes such as actin cytoskeleton regulation. The antiemetic drug domperidone, identified as an Hsp70 inhibitor, protected cells from TcdB (as well as TcdA and CDT) toxin intoxication by preventing the membrane translocation of TcdB's glucosyltransferase domain into the cytosol, revealing Hsp70 as a potential drug target for therapeutic strategies against severe *Clostridioides difficile* infections.

## 1. Introduction

*Clostridioides (C.) difficile*, formerly known as *Clostridium difficile*, is a spore-forming bacterium that can cause severe gastrointestinal infections in humans. These *C. difficile* infections (CDI) cause mild to severe diarrhea that can develop into life-threatening conditions such as pseudomembranous colitis, colonic perforation or toxic megacolon [1]. CDI is one of the most common causes of healthcare associated diarrhea and is linked to more than 200,000 hospitalizations in the US alone, leading to a significant healthcare burden, causing morbidity, mortality and increased healthcare costs [2]. Infections mainly occur if the gut microbiome has been disrupted, typically by previous antibiotic treatment. This allows *C. difficile* spores to develop to their vegetative form and then grow and secrete highly potent AB-type protein toxins, toxin A (TcdA) and toxin B (TcdB), which are the causative agents and main virulence factors for CDI symptoms [3]. TcdA and TcdB are encoded by the tcdA and tcdB genes, respectively, and are expressed in response to environmental cues such as exposure to bile acids [4]. Hypervirulent strains of *C. difficile* have distinct characteristics compared to other strains. They produce higher levels of TcdA and TcdB and exhibit greater resistance to antibiotics such as fluoroquinolones. In addition, these strains also produce a third toxin, the *C. difficile* transferase CDT, an actin ADP-ribosylating toxin that leads to the destruction of the actin cytoskeleton. Hypervirulent strains are associated with more severe disease, higher recurrence rates and enhanced mortality [3,5].

TcdA and TcdB are large (~300 kDa) glucosylating toxins that share about 50% sequence identity and act as AB-type toxins in the cytosol of target cells. Due to their complex structure, TcdA and TcdB are further classified as ABCD-type toxins. They consist of four functional domains: an enzymatically active glucosyltransferase domain (GTD) at the N-terminus (A-domain), an autoprotease domain (APD) (C-domain), a translocation domain that forms pores into endosomal compartments (D-domain) and a binding domain for receptor-binding of the toxin at the cell surface (B-domain). After receptor-binding, the toxins are internalized by receptor-mediated endocytosis. The acidification of endosomes leads to conformational changes in the toxins resulting in the formation of a translocation pore into the endosomal membrane [6]. The APD and GTD are translocated through this pore to the cytosol [7,8]. Cytosolic inositol hexakisphosphate ($InsP_6$) triggers the autoproteolytic activity of the APD leading to release of the GTD from the holotoxin [9,10]. The GTD glucosylates GTPases of the Rho-/Ras-family such as Cdc42, Rho or Rac1, which are located at the cell membrane [11,12]. They regulate numerous important functions and processes of cells such as cell migration, epithelial barrier integrity and cytokine production and are therefore referred to as molecular switches. Glucosylation inactivates these GTPases leading to release of certain cytokines and destruction of the cytoskeleton which in turn results in rounding of cells, loss of cell–cell contacts, cell cycle arrest and cell death [13]. In vivo, the inactivation of GTPases by the toxins is the major cause for gut barrier disruption, intestinal damage and appearance of clinical symptoms [2].

The treatment of CDI involves specific antibiotics including metronidazole, vancomycin and fidaxomicin, with the last two being superior to metronidazole. A particular problem is recurrent CDI, which is difficult to combat, and the risk of multiple relapses increases significantly with the first recurrence [2,14]. Since *C. difficile* toxins have been identified as one risk factor for recurrence, toxin-directed strategies take on greater significance. Bezlotoxumab, a human monoclonal antibody against TcdB, is in clinical use to reduce the risk of CDI recurrence [15].

In recent years, the role of host cell factors such as heat shock proteins and other chaperones during the translocation of bacterial AB-type toxins has been characterized (for a review, see [16]). Amongst others, clostridial binary toxins such as *Clostridium botulinum* C2 toxin and *C. difficile* CDT or *Bordetella pertussis* toxin and *Vibrio cholera* toxin require the activity of Hsp90, Hsp70, cyclophilins (Cyps) or FK506-binding proteins (FKBPs) for uptake of their enzyme subunits into the cytosol of target cells [17–30]. Interestingly, the intoxication of cells with TcdA or TcdB was independent of Hsp90, Cyps and FKBPs [17,19,25,31,32]. In the present study, we uncovered that Hsp70 activity is required for translocation of the GTD of TcdB into the cytosol and therefore for successful intoxication of cells with TcdB.

For characterizing the role of Hsp70 during TcdB uptake, we used specific pharmacological inhibitors of Hsp70 activity. VER-155008 (VER) is an established Hsp70 inhibitor that binds to the ATP binding site of Hsp70 [33]. ATP-binding and -hydrolysis is crucial for the function of Hsp70 during protein folding, translocation of proteins across membranes or refolding of misfolded proteins [34]. VER was previously used to characterize the role of Hsp70 during the uptake of other AB-type toxins such as pertussis toxin, diphtheria toxin or cholera toxin [24,26,35]. Recently, domperidone was discovered as a novel inhibitor of Hsp70 activity by a bioinformatics search for FDA-approved drugs that are structurally similar to the allosteric Hsp70 inhibitor HS-72, which also affects the ATP-binding site of Hsp70 [36]. Since the treatment of CDI is still challenging and novel therapeutic strategies focusing on the targeted pharmacological inhibition of the activity of the disease-causing toxins must be extended, we investigated the potential of the licensed drug domperidone (Dom) to protect human cells from TcdB intoxication and characterized the underlying inhibitory mechanism.

## 2. Results

### 2.1. Pharmacological Inhibition of Hsp70 Reduces the TcdB Induced Cell Rounding in a Time- and Concentration-Dependent Manner

Both Hsp70 inhibitors, VER and Dom, markedly delayed the intoxication of cells with TcdB in a concentration-dependent manner (Figure 1). The rounding of adherent cells such as HeLa or Vero cells results from the intracellular modification of Rho proteins by TcdB. Therefore, cell rounding represents a highly specific endpoint of intoxication indicating cytosolic uptake of the enzyme domain of TcdB. Figure 1a depicts that after 2 h of intoxication, HeLa cells treated solely with TcdB display a high number of rounded cells. If cells were treated with Dom or VER prior to TcdB intoxication, the number of rounded cells was reduced in a concentration-dependent manner. Quantification of rounded cells confirmed the inhibitory effect of VER and Dom on TcdB intoxication after 2 h of TcdB treatment (Figure 1b left). The delay of TcdB intoxication was visible for up to 8 h after intoxication (Figure 1b right). An $IC_{50}$ value for inhibition of TcdB by Dom of approximately 19 µM was determined (Figure 1c). The treatment of cells with Dom alone only showed a slight adverse effect in the highest concentration of 75 µM compared to untreated control samples (Figure 1d). Therefore, a maximal concentration of 50 µM of Dom was used in most of the following experiments. VER and Dom also protected Vero cells from TcdB intoxication (Figure 1e,f) and inhibited intoxication with TcdA (Figure 2).

The protective effect of Dom and VER on TcdB intoxication was also shown in the pathophysiologically more relevant human colon carcinoma cell line CaCo-2 (Figure 3a). CaCo-2 cells grow as a tight monolayer on transwell filters and serve as a physiologically relevant model of the intestinal epithelial barrier. TcdB causes a disruption of the epithelial integrity resulting in a decrease in transepithelial electrical resistance (TEER) (Figure 3b). In the presence of Dom, this TcdB-mediated decrease in resistance was delayed.

### 2.2. Pharmacological Inhibition of Hsp70 Leads to Reduced Glucosylation of Intracellular Rac1 by TcdB

To confirm the inhibitory effect of VER and Dom against TcdB, the glucosylation status of intracellular Rac1 was determined as an additional endpoint. After cells were intoxicated with TcdB in the presence or absence of different Dom/VER concentrations, lysates of these cells were analyzed by Western blot. A specific antibody was used that only recognizes the unmodified, i.e., non-glucosylated Rac1. Figure 4 shows that lysates from untreated control samples display a strong signal indicating that Rac1 was not glucosylated. Inhibitors alone had no significant effect on detection of unmodified Rac1. In cells treated with TcdB, no unmodified Rac1 was detectable, showing that most of the Rac1 was glucosylated in these cells. In the presence of Dom, a signal for non-glucosylated Rac1 was visible that was stronger than signals from samples treated with TcdB only (Figure 4). VER also inhibited Rac1 glucosylation by TcdB in higher concentrations, but the inhibitory effect of Dom was more pronounced. Therefore, we focused on characterizing the inhibitory effect of Dom in the following experiments.

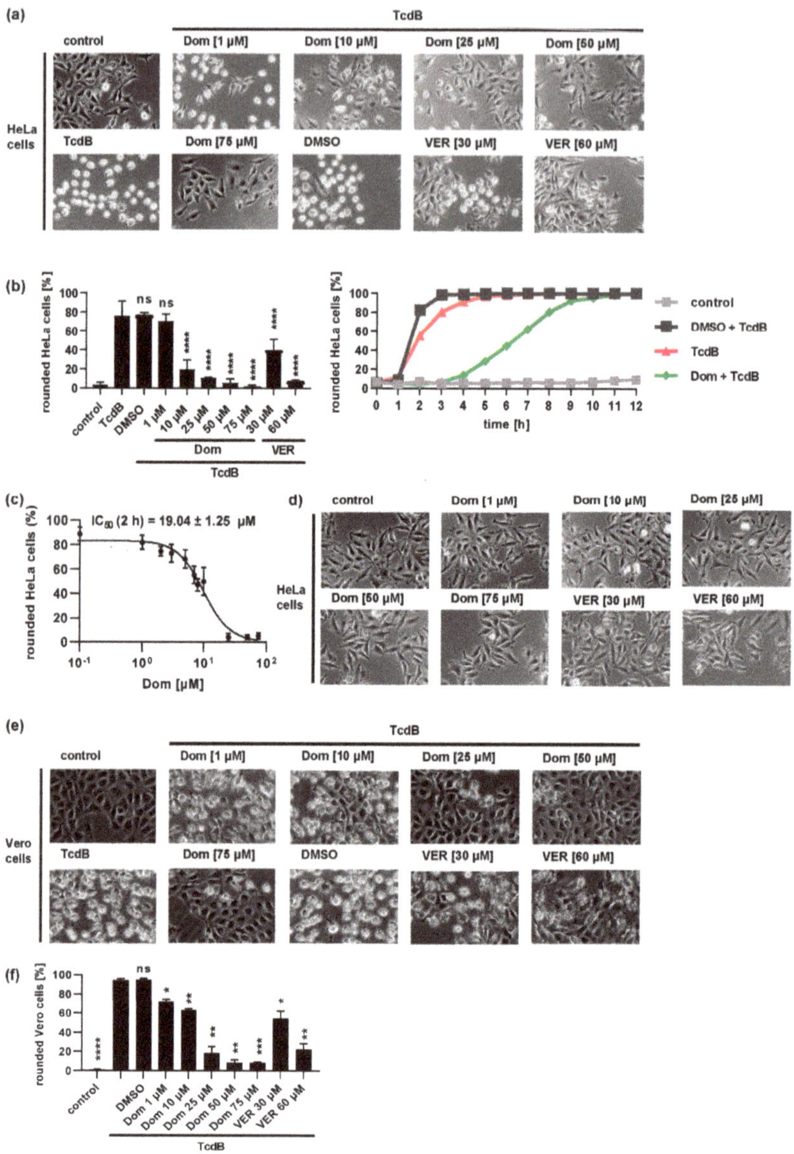


**Figure 1.** Hsp70 inhibition protects cells from TcdB intoxication ((**a**–**d**) HeLa cells; (**e**,**f**) Vero cells). (**a**) HeLa cells were pre-incubated with domperidone (Dom) or VER for 30 min at 37 °C. For control, cells were treated with DMSO as solvent control or left untreated. TcdB (30 pM) was added, or cells were left untreated (control). Images were obtained after 2 h of intoxication. (**b**) The percentage of rounded cells was determined from cell images shown in (**a**). Data after 2 h of intoxication (left, mean ± SD, n ≥ 3 of one representative of 3 independent experiments) and a time course of intoxication (right) are shown. For the time course, cells were pre-treated with 50 µM Dom and intoxicated with 30 pM TcdB. Values are given as percentage of control; single values per sample of 1 representative of 5 independent experiments are shown. Significance was tested by one-way ANOVA followed by Dunnett's multiple comparison test. All samples were compared to samples treated with TcdB only. (**c**) The concentration-dependent inhibition of TcdB (30 pM) was investigated by incubating HeLa cells with a concentrations series of Dom ranging from 0.1 µM to 75 µM. After 2 h, images were taken,

and percentage of rounded cells was determined. Values are given as mean ± SD (n ≥ 3 values of one representative of 3 independent experiments). Nonlinear fit was applied with GraphPad Prism software via log(inhibitor) vs. response (variable slope, four parameters). Under these conditions, an estimated $IC_{50}$ value of 19.04 ± 1.25 μM was calculated. (**d**) HeLa cells were treated with Dom or VER as described in (**a**). No toxin was added. Images were obtained after 2.5 h of inhibitor treatment. (**e**) Vero cells were treated with Dom, VER or DMSO or left untreated for control for 30 min. TcdB (30 pM) was added, or cells were left untreated (control). Images were obtained after 2 h of intoxication. (**f**) Percentage of rounded cells was determined from cell images shown in (**e**). Values are given as mean ± SD, n ≥ 3 (results of one representative of 3 independent experiments are shown). Significance was tested by one-way mixed-effects analysis followed by Dunnett's multiple comparison test. All samples were compared to samples treated with TcdB only (**** $p < 0.0001$, *** $p < 0.001$, ** $p < 0.01$, * $p < 0.05$, ns = not significant $p > 0.05$).

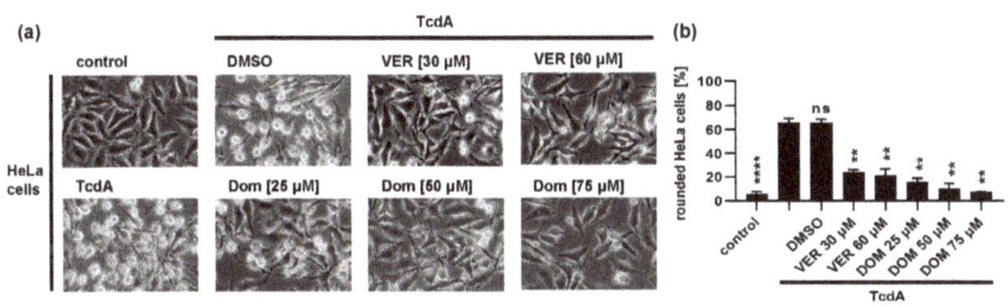


**Figure 2.** Dom and VER inhibit intoxication of HeLa cells with TcdA. (**a**) HeLa cells were pre-incubated with domperidone (Dom) or VER for 30 min at 37 °C. For control, cells were treated with DMSO as solvent control or left untreated. TcdA (10 nM) was added, or cells were left untreated (control). Images were obtained after 8 h of intoxication. (**b**) Percentage of rounded cells was determined from cell images shown in (**a**). Values are given as mean ± SD, n ≥ 3 (results of one of 3 independent experiments are shown). Significance was tested by mixed-effects analysis followed by Dunnett's multiple comparison test. All samples were compared to samples treated with TcdA only. (**** $p < 0.0001$, ** $p < 0.01$, ns = not significant $p > 0.05$).

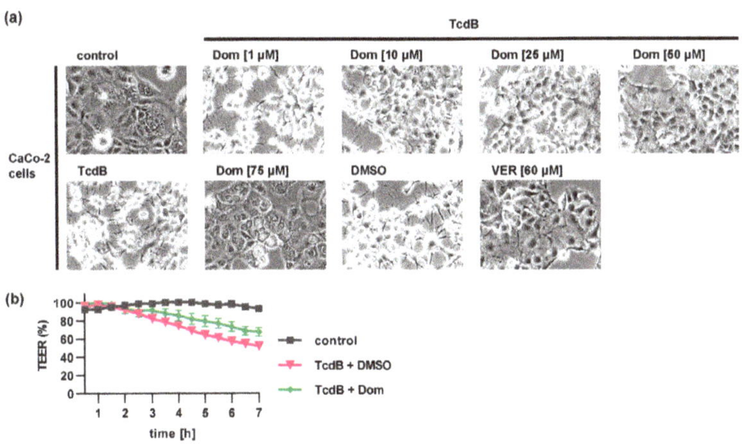


**Figure 3.** Hsp70 inhibition reduces the intoxication of CaCo-2 cells with TcdB. (**a**) CaCo-2 cells were pre-incubated with domperidone (Dom) or VER for 30 min at 37 °C. For control, cells were treated with DMSO as solvent control or left untreated. TcdB (30 pM) was added, or cells were left untreated (control). Images were obtained after 2 h of intoxication. (**b**) CaCo-2 cells were seeded in culture inserts

and treated with 50 μM Dom or DMSO for 30 min from the basolateral side. Then, TcdB (100 pM) was added apically, or cells were left untreated (control). TEER was measured every 30 min. Data were normalized to 0 h time point and are given as mean ± SEM, n ≥ 8 (from 6 independent experiments).

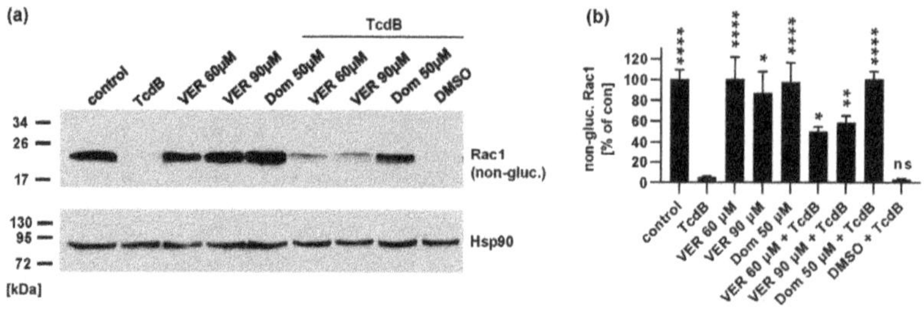


**Figure 4.** Domperidone inhibits TcdB-mediated glucosylation of intracellular Rac1. HeLa cells were pre-incubated with domperidone (Dom) or VER for 30 min at 37 °C. For control, cells were treated with DMSO as solvent control or left untreated. TcdB (30 pM) was added, or cells were left untreated (control). After 2 h, cells were harvested. Lysates were analyzed by SDS-PAGE and Western blotting. Non-glucosylated Rac1 signal as well as Hsp90 signal for loading control were detected. (**a**) A representative Western blot is shown. (**b**) Western blot signals were quantified and normalized to Hsp90 signals as well as to untreated control samples. Values are given as mean ± SEM (n ≥ 4, of at least 2 independent experiments). Significance was tested by one-way ANOVA analysis and multiple comparisons test and refers to samples treated with only TcdB. (**** $p < 0.0001$, ** $p < 0.01$, * $p < 0.05$, ns = not significant $p > 0.05$).

### 2.3. *Domperidone Has no Inhibitory Effect on TcdB Receptor Binding, Autoprotease Activity or In Vitro Glucosyltransferase Activity*

To unravel the underlying mechanism of inhibition, it was next tested whether Dom inhibits crucial steps of TcdB uptake and its mode of action. Dom had no inhibitory effect on binding of TcdB to cells (Figure 5a). For releasing the enzyme domain of TcdB into the cytosol, the autoproteolytic activity is required. The incubation of TcdB with $InsP_6$ triggers autoproteolysis. Figure 5b shows that in the presence of different Dom concentrations, TcdB is still autoproteolytically processed. For control, N-ethylmaleimide (NEM) was included as a known inhibitor of autoproteolysis [10]. Glucosyltransferase activity of TcdB in vitro was not inhibited by Dom, as demonstrated by the UDP-Glo$^{TM}$ glucosyltransferase assay (Figure 5c) and by the detection of non-glucosylated Rac1 via Western blot (Figure 5d).

### 2.4. *Domperidone Inhibits pH-Driven Membrane Translocation of the Enzyme Domain of TcdB*

Since Dom had no effect on the receptor binding, autoproteolytic activity and enzyme activity, another step of TcdB uptake must be affected by Dom. It was already shown before that Hsp70 facilitates the translocation of other AB-type toxins across endosomal membranes. Therefore, we next investigated if Dom inhibits TcdB membrane translocation in an isolated manner. To achieve this, we aimed to replicate the acidic conditions found in endosomes directly at the cytoplasmic membrane as described before [8]. By doing so, we could induce translocation of the enzyme domain across the cytoplasmic membrane without relying on the endocytosis route and subsequent translocation from endosomes. To initiate this process, cells were incubated with TcdB while kept on ice to facilitate binding while prevent internalization. Challenging cells with warm acidic medium triggers pore formation by TcdB directly into the cytoplasmic membrane and translocation of the glucosyltransferase domain into the cytosol. To block the acidification of endosomes and therefore the physiological uptake route of TcdB, cells were also treated with bafilomycinA1 (BafA1). BafA1 inhibits the vesicular ATPase which pumps protons into the endosome. Pre-incubation of cells with Dom prevented membrane translocation of TcdB (Figure 6a). If Dom was only present during the acidification step, inhibition was not observed (Figure 6b).

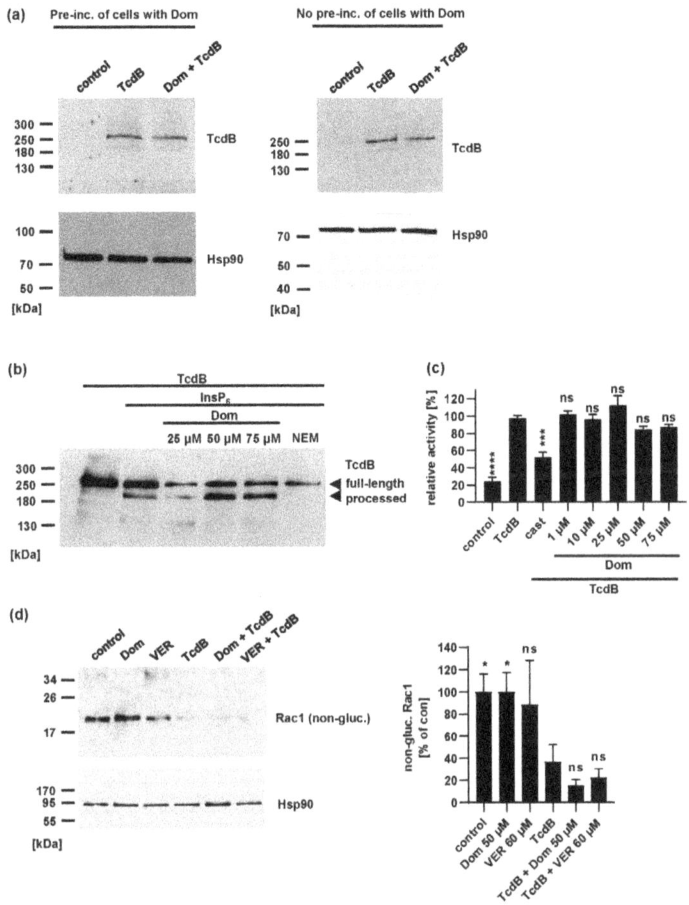


**Figure 5.** Domperidone has no inhibitory effect on receptor binding, autoproteolysis or enzyme activity of TcdB. (**a**) Vero cells were treated with 50 μM domperidone (Dom), either with (left) or without (right) 30 min pre-incubation at 37 °C, before TcdB (500 pM) was added for 1 h on ice. TcdB and Hsp90 (loading control) were detected by Western blotting using specific antibodies. pre-inc. = pre-incubation (**b**) For assessing the effect of Dom on the autoproteolytic activity of TcdB, 2 μg TcdB was incubated with indicated concentrations of Dom plus inositol hexakisphosphate ($InsP_6$). For control, TcdB was treated with $InsP_6$ alone or left untreated. N-ethylmaleimide (NEM) was included as a known inhibitor of TcdB autoproteolytic activity. Samples were incubated for 1 h at 37 °C and subsequently subjected to SDS-PAGE and Western blotting. Full-length TcdB and processed fragments were detected using a specific antibody. (**c**) UDP-Glo$^{TM}$ glucosyltransferase assay was performed to determine the effect of Dom on TcdB glucosyltransferase activity. A total of 200 pM TcdB were incubated with indicated concentrations of Dom in the presence of Rac1. Castanospermin (cast) was included as a known inhibitor of TcdB glucosyltransferase activity. Via the addition of UDP-glucose, the reaction was initiated and proceeded for 1 h at 37 °C. The UDP detection reagent was added, and luminescence was measured. Values were normalized to samples treated only with TcdB and are given as mean ± SEM, n ≥ 13 values from 5 independent experiments. Outliers were removed according to GraphPad Prism outlier test. Significance was tested by mixed-effects analysis and multiple comparisons test and refer to samples treated only with TcdB. (**d**) 40 μg cell lysate was incubated with 50 ng TcdB with or without Dom or VER for 2 h at 37 °C. For control, lysate was incubated

with VER or Dom alone. Non-glucosylated Rac1 was detected by Western blot analysis. A representative blot (left) and quantification of Western blot signals (right) are shown. Quantified signals were normalized to Hsp90 signals and to untreated control. Values are given as mean ± SEM, n ≥ 6 values from 4 independent experiments. Significance was tested against samples treated only with TcdB by mixed-effects analysis and multiple comparisons test. (**** $p < 0.0001$, *** $p < 0.001$, * $p < 0.05$, ns = not significant $p > 0.05$).

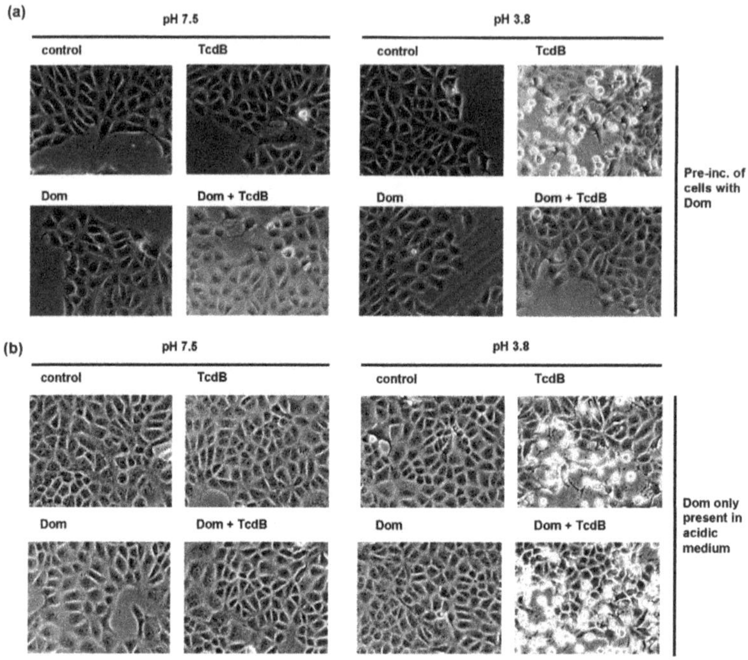


**Figure 6.** Domperidone inhibits membrane translocation of TcdB. (**a**) Vero cells were pre-incubated with domperidone (Dom, 50 μM) for 30 min or left untreated. All samples were treated with bafilomycinA1 (BafA1, 100 nM) for another 30 min. Cells were incubated on ice for 10 min, followed by addition of TcdB (500 pM) for 30 min on ice. For control, cells were left untreated (control). Then, warm acidic medium containing 50 μM Dom was added for 10 min at 37 °C. For control, warm neutral medium with 50 μM Dom was added. Subsequently, medium was removed and replaced with BafA1-containing medium. Morphological changes were monitored, and images show cells after 6 h of intoxication. (**b**) Vero cells were incubated with BafA1 (100 nM) for 30 min. Then, cells were put on ice for 10 min, followed by addition of TcdB (500 pM) for 30 min on ice. For control, cells were left untreated (control). Then, warm acidic medium containing 50 μM Dom was added for 10 min at 37 °C. For control, warm neutral medium with 50 μM Dom was added. Afterwards, medium was replaced with BafA1-containing medium, and morphological changes were monitored. Images show cells after 5 h of intoxication. Pre-inc. = pre-incubation.

### *2.5. Domperidone Reduces Intoxication of Cells with Binary CDT*

Hypervirulent *C. difficile* strains produce the ADP-ribosylating CDT toxin in addition to TcdB and TcdA. CDT ADP-ribosylates G-actin leading to depolymerization of F-actin and rounding of adherent cells (Figure 7). Cells treated with increasing concentrations of Dom and CDT showed fewer rounded cells compared to cells treated with CDT alone indicating inhibition of intoxication (Figure 7a). Quantification of rounded cells confirmed the inhibitory effect of Dom on CDT intoxication of cells (Figure 7b,c).

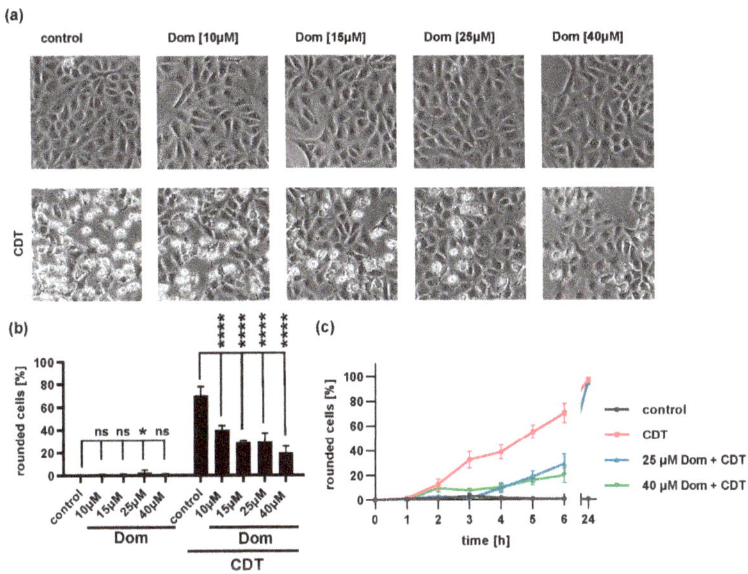


**Figure 7.** Domperidone protects cells from CDT intoxication. Vero cells were pre-incubated with indicated concentrations of domperidone (Dom) for 30 min at 37 °C or left untreated for control. Then, cells were challenged with CDT (150 ng/mL CDTa plus 300 ng/mL CDTb), or cells were further incubated with the inhibitor alone. Pictures were taken every hour. (**a**) Morphological changes after 6 h of CDT intoxication are shown. (**b**) The percentage of rounded cells was determined after 6 h of CDT intoxication from images. Values are given as mean ± SD, n = 3 values of one representative experiment. Significance was tested by one-way ANOVA and Dunnett's multiple comparisons test. (**c**) A time-course of intoxication with CDT is shown. Values are given as mean ± SD, n = 3 values of one representative experiment. (**** $p < 0.0001$, * $p < 0.05$, ns = not significant $p > 0.05$).

Taken together, the results indicate that Hsp70 inhibitors Dom and VER reduced the intoxication of cells with TcdB by analyzing TcdB-induced cell morphology as well as the glucosylation status of intracellular TcdB substrate Rac1 and TcdB-mediated decrease in resistance of CaCo-2 monolayers. Dom had no effect on TcdB cell binding, autoprotease activity or enzyme activity in vitro but inhibited membrane translocation of the glucosyltransferase domain of TcdB into the cytosol of target cells. Moreover, the intoxication of cells with TcdA as well as the binary CDT toxin, produced by hypervirulent strains in addition to glucosylating toxins, was also reduced by Dom.

## 3. Discussion

Hsp70 is a highly conserved, abundant protein that is part of the molecular chaperone family and works as a co-chaperone of Hsp90. It plays a role in various cellular processes, including protein folding, translocation of polypeptide chains into the mitochondria or endoplasmic reticulum, and preventing protein aggregation and misfolding. Hsp70 helps in protein translocation by binding the protein close to the translocation pore and generating a one-way pulling motion, known as entropic pulling [34,37,38]. Hsp70 comprises four domains: the nucleotide binding domain, substrate binding domain, helical lid domain and C-terminal tail domain [34]. ATP binding and hydrolysis are vital for Hsp70 function, and inhibitors such as VER-155008 (VER) prevent this process [33].

The requirement of Hsp70 for membrane translocation has been shown for other AB-type toxins before (for review see [16]). ADP-ribosylating toxins including clostridial binary toxins C2 toxin, iota toxin and CDT toxin [21,22], diphtheria toxin [35], pertussis toxin [24] and cholera toxin [26] depend on Hsp70 activity for the uptake of their enzyme subunits into the cytosol of host cells and therefore for intoxication of cells. These toxins also rely on the activity of additional cellular chaperones, namely, Hsp90, cyclophilins (Cyps) and FK506 binding proteins (FKBPs) [16]. During these investigations, it was shown that the uptake of TcdB occurred without dependence on Hsp90, while the uptake of TcdA occurred without reliance on Cyps and FKBPs [17,19,25]. Given the close structural and functional relationship between TcdA and TcdB, it is highly probable that both toxins operate independently of Hsp90, Cyps and FKBPs but require Hsp70 activity. These findings point towards a common necessity for chaperones in uptake of AB-type toxins. However, it is notable that the specific types of chaperones required can differ depending on the toxin in question.

Later, it was shown that TcdB and TcdA need the assistance of the chaperonin TRiC/CCT (TCP-1 ring complex, chaperonin containing TCP-1) for efficient refolding and restoration of the glucosyltransferase activity of the enzyme domain [32]. TRiC/CCT is a ring-shaped molecular machine involved in the folding of many newly synthesized proteins. It appears that approximately 10% of cytosolic proteins interact with TRiC/CCT [39,40]. While the CCT4/5 subunits of TRiC/CCT have been shown to be effective in refolding of the TcdB enzyme domain (TcdB$^{GT}$), it is not known if they are involved in membrane translocation [32]. Notably, the enzyme activity of TcdB$^{GT}$ was not recovered by Hsp90, and only a weak recovery effect was seen with Hsp70 [32]. However, our findings clearly demonstrate that Dom effectively inhibits membrane translocation of TcdB, suggesting that TRiC/CCT and Hsp70 perform complementary functions. Hsp70 aids in the translocation from the endosome to the cytosol, possibly through entropic pulling that has been described for Hsp70 before, while TRiC/CCT takes over refolding and restoration of TcdB$^{GT}$ enzyme activity.

Recent research has revealed that Dom has an inhibitory effect on Hsp70 by screening a library of FDA-approved drugs for compounds with structural similarity to the known Hsp70 inhibitor HS-72 [36,41]. In a study conducted by Concilli et al. in 2020, it was shown that Dom and HS-72 both significantly reduced the ATPase activity of Hsp70, with Dom exhibiting an even stronger effect than HS-72 [36]. Dom is widely used in many countries to treat nausea and vomiting associated with conditions such as gastroesophageal reflux disease (GERD) or gastroparesis. Adverse effects include headache, dry mouth, dizziness, and gastrointestinal disturbances. In rare cases, serious adverse effects such as cardiac arrhythmias were reported [42,43].

The treatment of CDI typically involves the use of specific antibiotics that target the bacteria causing the infection, such as metronidazole, vancomycin and fidaxomicin [1]. However, treatment is particularly challenging in cases of recurrent CDI. Bezlotoxumab is a monoclonal antibody approved by the FDA in 2016 for use in combination with standard antibiotics for the treatment of CDI in adults who are at high risk of recurrence [44]. It works by binding to and specifically neutralizing TcdB, which can cause severe inflammation and damage to the intestinal lining. In combination with standard antibiotic therapy, bezlotoxumab reduces the rate of CDI recurrence. Therefore, bezlotoxumab is considered an effective treatment option for reducing the risk of recurrence in patients with *C. difficile* infection [15,44].

Bezlotoxumab is the only approved drug targeting a *C. difficile* toxin and only neutralizes TcdB that has not yet bound to cells [44]. Since TcdA and TcdB cause the severe symptoms during the infection, new therapeutic strategies should focus on targeting these toxins. In this regard, inhibiting Hsp70 to prevent translocation of the enzyme domain into the cytosol represents an opportunity to neutralize toxin molecules that have already been internalized. Domperidone, as an approved drug with a known safety profile, may offer an option to extend the therapeutic approach to treat CDI. However, careful consideration of therapeutic efficacy versus adverse effects, depending on the severity of the disease, is necessary.

Hypervirulent strains of *C. difficile* are associated with more severe disease, higher rates of recurrence and increased mortality. They produce the binary CDT toxin in addition to TcdA and TcdB. Previous studies have shown that the ADP-ribosylating CDT toxin also requires Hsp70 as well as Hsp90, Cyp and FKBP activity for membrane translocation of its enzyme component into the cytosol, making Dom a potentially interesting treatment option for infections caused by hypervirulent *C. difficile* strains [21,22,45,46]. Additionally, other toxins such as pertussis toxin, diphtheria toxin and cholera toxin, which also cause severe diseases, depend on Hsp70 for their action. Therefore, Dom treatment might have therapeutic potential for treatment of diseases caused by these bacterial toxins, including whooping cough, diphtheria or cholera.

## 4. Materials and Methods

### *4.1. Protein Expression and Purification*

Toxin components CDTa and CDTb were expressed, purified and activated as recombinant proteins as described before [47]. Native TcdB from *C. difficile* VPI 10,463 was expressed and purified as described earlier [48]. Rac1 was expressed and purified as a recombinant GST-tagged protein as described previously [49]. Native TcdA was obtained from List (Campbell, CA, USA).

### *4.2. Cell Culture and Intoxication Experiments*

HeLa (DSMZ, Braunschweig, Germany) and Vero cells (African green monkey kidney cells, DSMZ, Braunschweig, Germany) were cultivated in minimum essential medium (MEM) with 10% heat-inactivated fetal calf serum (FCS) from GIBCO life technologies (Karlsruhe, Germany), 1 mM sodium pyruvate (Thermo Fisher Scientific, Waltham, MA, USA), 2 mM L-glutamine (PAN-Biotech, Aidenbach, Germany), 0.1 mM non-essential amino acids, 100 U/mL of penicillin and 100 µg/mL of streptomycin (Thermo Fisher Scientific).

CaCo-2 cells (human epithelial colorectal adenocarcinoma cells, ATCC HTB-37, Manassas, VA, USA) were cultured in Dulbecco's Modified Eagle Medium (DMEM) from GIBCO life technologies with 10% FCS, 1 mM sodium pyruvate, 0.1 mM non-essential amino acids, 100 U/mL penicillin and 100 µg/mL streptomycin.

Humidified conditions with 5% $CO_2$ at 37 °C were applied to incubate cells. For intoxication experiments, cells were detached with trypsin and seeded into 24-well plates (TPP Techno Plastic Products, Trasadingen, Switzerland). Medium was exchanged, and cells were pre-incubated with domperidone (Sigma Aldrich by Merck, Darmstadt, Germany), VER (inhibitor of ATP- binding site of Hsp70, Hsc70 and Grp78, Tocris Bioscience, Wiesbaden-Nordenstadt, Germany), BafA1 (Calbiochem, Bad Soden, Germany), DMSO as solvent control or medium only for 30 min. Then, toxin was added, and cell images were taken after different time points using a Zeiss (Oberkochen, Germany) Axiovert 40CFL microscope with a Jenoptik (Jena, Germany) ProGres C10 CCD camera/Leica DMi1 microscope connected to a Leica MC170 HD camera (both Leica Microsystems GmbH, Wetzlar, Germany). The percentage of rounded, i.e., intoxicated, cells was determined from images using neuralab.de.

To obtain a time course of intoxication, the zenCELL owl 24 channel automated microscope (innoME GmbH, Espelkamp, Germany) was used. Images were obtained every 10 min for 8–12 h.

### *4.3. Glucosylation Status of Intracellular Rac1 after TcdB-Treatment*

Cells were pre-incubated with the respective inhibitors and treated with TcdB as described above. After 2 h, cells were washed and lysed by freezing. Cell lysates here harvested in 2.5 Laemmli + DTT. Samples were subjected to SDS-PAGE followed by Western blotting. Non-glucosylated Rac1 (1:1000, #610651, BD Biosciences, Franklin Lakes, NJ, USA) as well as Hsp90 (1:1000, Santa Cruz Biotechnology, Heidelberg, Germany) were detected by specific primary antibodies and horseradish peroxidase-coupled secondary antibodies (Santa Cruz Biotechnology) using the enhanced chemiluminescence system (ECL, Millipore, Merck, Darmstadt, Germany). Densitometric quantification of Western blot signals was determined using ImageJ (v1.53p, U.S. National Institutes of Health, Bethesda, MD, USA). Values were normalized to loading control signals (Hsp90).

### 4.4. Cell-Binding Assay

Cells were treated with Dom, either with or without a 30 min pre-incubation. Then, cells were transferred to ice for 10 min before TcdB was added for 1 h on ice. Samples were subjected to SDS-PAGE and Western blotting. TcdB (1:1000, MCA4737, BioRad, Hercules, CA, USA) and Hsp90 were detected by specific antibodies.

### 4.5. In Vitro Autoprocessing of TcdB

Effects of Dom on the intrinsic cysteine protease activity of TcdB were tested by incubating TcdB with or without different concentrations of Dom for 1 h at 37 °C in a 20 mM Tris-HCl buffer containing 150 mM NaCl at pH 7.4. To induce autoprocessing activity, 1 mM inositol hexakisphosphate (Santa Cruz Biotechnology) was added. As a positive control for inhibition of autoprocessing, 1 mM N-ethylmaleimide (Sigma Aldrich by Merck) was used. To stop the reaction, Laemmli buffer was added, and samples were denatured at 95 °C. Samples were subjected to SDS-PAGE (8% gel) followed by Western blotting. TcdB and fragments were detected by a specific primary antibody (1:1000, MCA4737, BioRad).

### 4.6. UDP-Glo$^{TM}$ Glucosylation Assay

The assay was performed according to the manufacturer's protocol (Promega, Madison, WI, USA). A total of 200 pM TcdB were incubated with indicated concentrations of Dom in glucosylation buffer (50 mM HEPES, 100 mM KCl, 2 mM $MgCl_2$, 1 mM $MnCl_2$, 100 mg/L BSA, pH 7.5). TcdB substrate, recombinant Rac1 (5 µM), was added, and the reaction was started by addition of 100 µM UDP-glucose for 1 h at 37 °C. For control, the inhibitor of TcdB glucosyltransferase activity, castanospermin (10 mM, Santa Cruz Biotechnology), was included [50]. Three 10 µL aliquots of each sample were transferred to a white half area 96-well plate (Greiner Bio-One International GmbH, Kremsmünster, Austria) and treated with 10 µL of UDP detection reagent for 30 s on a plate shaker. Luminescence was measured after 15–60 min by Tecan infinite M1000 Pro plate reader (Tecan, Männsdorf, Switzerland) with 750 ms integration time.

### 4.7. In Vitro Glucosylation of Rac1 by TcdB

In this stage, 40 µg cell lysate was incubated with 50 ng TcdB in glucosylation buffer (50 mM HEPES, 100 mM KCl, 2 mM $MgCl_2$, 1 mM $MnCl_2$, 100 mg/L BSA, pH 7.5) in the presence of Dom or VER at indicated concentrations for 2 h at 37 °C. Samples were analyzed by SDS-PAGE and Western blotting. Non-glucosylated Rac1 was detected by a specific antibody (1:500, #61065, BD Biosciences, Franklin Lakes, NJ, USA). Hsp90 was detected to confirm equal loading. Signals were detected using the ECL system. Signals were densitometrically quantified using ImageJ. Non-glucosylated Rac1 signals were normalized to Hsp90 signals and then to untreated control samples.

### 4.8. Toxin Translocation Assay

The assay was performed as described earlier [51]. In brief, cells were pre-incubated with the respective inhibitor for 30 min followed by another 30 min incubation with medium containing the inhibitor plus BafA1. BafA1, an inhibitor of the vesicular ATPase, prevents the normal toxin uptake route via escape from endosomes. Subsequently, cells were incubated on ice with TcdB for 30 min. Then, cells were challenged with warm acidic medium (pH 3.8) for 10 min at 37 °C triggering toxin pore formation as well as translocation of the enzyme domain across the cytoplasmic membrane directly into the cell cytosol. For control, cells were treated with neutral medium (pH 7.5). Then, medium was exchanged for fresh BafA1-containing medium (pH 7.5). Morphological changes (cell rounding) were monitored and documented as a specific endpoint of intoxication.

### 4.9. Transepithelial Electrical Resistance (TEER) Measurements

CaCo-2 cells were seeded in 24-well culture inserts with 0.4 µm pore size (Corning Inc., Kennebunk ME, USA). Cells were grown for 3–5 days. After 30 min pre-incubation with the respective inhibitor, 100 pM TcdB were added, and cells were further incubated at 37 °C. Inhibitors were applied from the basolateral, and TcdB was applied from the apical side of the filter insert. For control, cells were left untreated or were treated with DMSO as solvent control for inhibitors. TEER was measured every hour with the EVOM2 apparatus using a STX2 electrode (World Precision Instruments Inc., Sarasota, FL, USA). Raw resistance data were normalized to 0 h time point.

### 4.10. Reproducibility of Experiments and Statistics

All experiments were performed independently at least 3 times. Number of replicates is given in the respective figures. Representative results are shown in the figures. Western blots were cropped for display reasons only. Data were statistically analyzed as stated in the figure legends (**** $p < 0.0001$, *** $p < 0.001$, ** $p < 0.01$, * $p < 0.05$, ns = not significant $p > 0.05$).

**Author Contributions:** Conceptualization, P.P., H.B. and K.E.; formal analysis, M.B.-Y., J.J., M.W., J.S. and K.E.; funding acquisition, H.B. and K.E.; investigation, M.B.-Y., J.J., M.W., J.S. and K.E.; project administration, K.E.; resources, H.B. and K.E.; supervision, K.E.; visualization, M.B.-Y., J.J., M.W. and K.E.; writing—original draft, K.E. All authors have read and agreed to the published version of the manuscript.

**Funding:** This work was supported by the Deutsche Forschungsgemeinschaft (DFG) GRK2203 Projektnummer 278012962 (grant to K.E.) and DFG—Projektnummer 450938962 (grant to H.B.).

**Institutional Review Board Statement:** Not applicable.

**Informed Consent Statement:** Not applicable.

**Data Availability Statement:** The datasets generated and/or analyzed during the current study are either included in the study or available from the corresponding author on reasonable request.

**Acknowledgments:** K.E. is a fellow of the Margarete von Wrangell program supported by the European Social Fund and Ministry of Science, Research and Art Baden-Württemberg. M.B.Y. is a fellow of the International Graduate School in Molecular Medicine Ulm (IGradU). We thank Klaus Aktories for providing toxins, Barbara Moepps and Michael Meister for providing the GST-Rac1 plasmid and Anna Anastasia for excellent technical assistance. J.J. is a fellow of the IGradU Nanjing program (medical faculty of Ulm University, Southeast University (SEU) Nanjing China, Chinese Scholarship Council (CSC)).

**Conflicts of Interest:** The authors declare no conflict of interest.

## References

1. Johnson, S.; Lavergne, V.; Skinner, A.M.; Gonzales-Luna, A.J.; Garey, K.W.; Kelly, C.P.; Wilcox, M.H. Clinical Practice Guideline by the Infectious Diseases Society of America (IDSA) and Society for Healthcare Epidemiology of America (SHEA): 2021 Focused Update Guidelines on Management of *Clostridioides difficile* Infection in Adults. *Clin. Infect. Dis.* **2021**, *73*, e1029–e1044. [CrossRef]
2. Kordus, S.L.; Thomas, A.K.; Lacy, D.B. *Clostridioides difficile* Toxins: Mechanisms of Action and Antitoxin Therapeutics. *Nat. Rev. Microbiol.* **2022**, *20*, 285–298. [CrossRef]
3. Kuehne, S.A.; Collery, M.M.; Kelly, M.L.; Cartman, S.T.; Cockayne, A.; Minton, N.P. Importance of Toxin A, Toxin B, and CDT in Virulence of an Epidemic Clostridium Difficile Strain. *J. Infect. Dis.* **2014**, *209*, 83–86. [CrossRef]
4. Shen, A. *Clostridioides difficile* Spores: Bile Acid Sensors and Trojan Horses of Transmission. *Clin. Colon Rectal Surg.* **2020**, *33*, 58–66. [CrossRef]
5. Abeyawardhane, D.L.; Godoy-Ruiz, R.; Adipietro, K.A.; Varney, K.M.; Rustandi, R.R.; Pozharski, E.; Weber, D.J. The Importance of Therapeutically Targeting the Binary Toxin from Clostridioides Difficile. *Int. J. Mol. Sci.* **2021**, *22*, 2926. [CrossRef]
6. Qa'Dan, M.; Spyres, L.M.; Ballard, J.D. PH-Induced Conformational Changes InClostridium Difficile Toxin B. *Infect. Immun.* **2000**, *68*, 2470–2474. [CrossRef]
7. Pfeifer, G.; Schirmer, J.; Leemhuis, J.; Busch, C.; Meyer, D.K.; Aktories, K.; Barth, H. Cellular Uptake of Clostridium Difficile Toxin B: Translocation of the N-terminal catalytic domain into the cytosol of eukaryotic cells. *J. Biol. Chem.* **2003**, *278*, 44535–44541. [CrossRef]

8. Barth, H.; Pfeifer, G.; Hofmann, F.; Maier, E.; Benz, R.; Aktories, K. Low PH-Induced Formation of Ion Channels by Clostridium Difficile Toxin B in Target Cells. *J. Biol. Chem.* **2001**, *276*, 10670–10676. [CrossRef]
9. Reineke, J.; Tenzer, S.; Rupnik, M.; Koschinski, A.; Hasselmayer, O.; Schrattenholz, A.; Schild, H.; von Eichel-Streiber, C. Autocatalytic Cleavage of Clostridium Difficile Toxin B. *Nature* **2007**, *446*, 415–419. [CrossRef]
10. Egerer, M.; Giesemann, T.; Jank, T.; Satchell, K.J.F.; Aktories, K. Auto-Catalytic Cleavage of Clostridium Difficile Toxins A and B Depends on Cysteine Protease Activity. *J. Biol. Chem.* **2007**, *282*, 25314–25321. [CrossRef]
11. Just, I.; Selzer, J.; Wilm, M.; Eichel-Streiber, C.; von Mann, M.; Aktories, K. Glucosylation of Rho Proteins by Clostridium Difficile Toxin B. *Nature* **1995**, *375*, 500–503. [CrossRef]
12. Just, I.; Wilm, M.; Selzer, J.; Rex, G.; von Eichel-Streiber, C.; Mann, M.; Aktories, K. The Enterotoxin from Clostridium Difficile (ToxA) Monoglucosylates the Rho Proteins. *J. Biol. Chem.* **1995**, *270*, 13932–13936. [CrossRef]
13. Etienne-Manneville, S.; Hall, A. Rho GTPases in Cell Biology. *Nature* **2002**, *420*, 629–635. [CrossRef]
14. Yokoyama, Y.; Shiota, A.; Asai, N.; Koizumi, Y.; Yamagishi, Y.; Sakanashi, D.; Nakamura, A.; Suematsu, H.; Ohnishi, M.; Mikamo, H. Risk Factors of First Recurrence of *Clostridioides difficile* Infection. *Anaerobe* **2022**, *75*, 102556. [CrossRef]
15. Wilcox, M.H.; Gerding, D.N.; Poxton, I.R.; Kelly, C.; Nathan, R.; Birch, T.; Cornely, O.A.; Rahav, G.; Bouza, E.; Lee, C.; et al. Bezlotoxumab for Prevention of Recurrent Clostridium Difficile Infection. *N. Engl. J. Med.* **2017**, *376*, 305–317. [CrossRef]
16. Ernst, K. Requirement of Peptidyl-Prolyl Cis/Trans Isomerases and Chaperones for Cellular Uptake of Bacterial AB-Type Toxins. *Front. Cell. Infect. Microbiol.* **2022**, *12*, 938015. [CrossRef]
17. Haug, G.; Leemhuis, J.; Tiemann, D.; Meyer, D.K.; Aktories, K.; Barth, H. The Host Cell Chaperone Hsp90 Is Essential for Translocation of the Binary Clostridium Botulinum C2 Toxin into the Cytosol. *J. Biol. Chem.* **2003**, *278*, 32266–32274. [CrossRef]
18. Kaiser, E.; Kroll, C.; Ernst, K.; Schwan, C.; Popoff, M.; Fischer, G.; Buchner, J.; Aktories, K.; Barth, H. Membrane Translocation of Binary Actin-ADP-Ribosylating Toxins from Clostridium Difficile and Clostridium Perfringens Is Facilitated by Cyclophilin A and Hsp90. *Infect. Immun.* **2011**, *79*, 3913–3921. [CrossRef]
19. Kaiser, E.; Böhm, N.; Ernst, K.; Langer, S.; Schwan, C.; Aktories, K.; Popoff, M.; Fischer, G.; Barth, H. FK506-Binding Protein 51 Interacts with Clostridium Botulinum C2 Toxin and FK506 Inhibits Membrane Translocation of the Toxin in Mammalian Cells. *Cell. Microbiol.* **2012**, *14*, 1193–1205. [CrossRef]
20. Ernst, K.; Langer, S.; Kaiser, E.; Osseforth, C.; Michaelis, J.; Popoff, M.R.; Schwan, C.; Aktories, K.; Kahlert, V.; Malesevic, M.; et al. Cyclophilin-Facilitated Membrane Translocation as Pharmacological Target to Prevent Intoxication of Mammalian Cells by Binary Clostridial Actin ADP-Ribosylated Toxins. *J. Mol. Biol.* **2015**, *427*, 1224–1238. [CrossRef]
21. Ernst, K.; Schmid, J.; Beck, M.; Hägele, M.; Hohwieler, M.; Hauff, P.; Ückert, A.K.; Anastasia, A.; Fauler, M.; Jank, T.; et al. Hsp70 Facilitates Trans-Membrane Transport of Bacterial ADP-Ribosylating Toxins into the Cytosol of Mammalian Cells. *Sci. Rep.* **2017**, *7*, 2724. [CrossRef] [PubMed]
22. Ernst, K.; Liebscher, M.; Mathea, S.; Granzhan, A.; Schmid, J.; Popoff, M.R.; Ihmels, H.; Barth, H.; Schiene-Fischer, C. A Novel Hsp70 Inhibitor Prevents Cell Intoxication with the Actin ADP-Ribosylating Clostridium Perfringens Iota Toxin. *Sci. Rep.* **2016**, *6*, 20301. [CrossRef] [PubMed]
23. Ernst, K.; Eberhardt, N.; Mittler, A.-K.; Sonnabend, M.; Anastasia, A.; Freisinger, S.; Schiene-Fischer, C.; Malešević, M.; Barth, H. Pharmacological Cyclophilin Inhibitors Prevent Intoxication of Mammalian Cells with *Bordetella pertussis* Toxin. *Toxins* **2018**, *10*, 181. [CrossRef] [PubMed]
24. Ernst, K.; Mittler, A.-K.; Winkelmann, V.; Kling, C.; Eberhardt, N.; Anastasia, A.; Sonnabend, M.; Lochbaum, R.; Wirsching, J.; Sakari, M.; et al. Pharmacological Targeting of Host Chaperones Protects from Pertussis Toxin in Vitro and in Vivo. *Sci. Rep.* **2021**, *11*, 5429. [CrossRef]
25. Kaiser, E.; Pust, S.; Kroll, C.; Barth, H. Cyclophilin A Facilitates Translocation of the Clostridium Botulinum C2 Toxin across Membranes of Acidified Endosomes into the Cytosol of Mammalian Cells. *Cell. Microbiol.* **2009**, *11*, 780–795. [CrossRef]
26. Burress, H.; Kellner, A.; Guyette, J.; Tatulian, S.A.; Teter, K. HSC70 and HSP90 Chaperones Perform Complementary Roles in Translocation of the Cholera Toxin A1 Subunit from the Endoplasmic Reticulum to the Cytosol. *J. Biol. Chem.* **2019**, *294*, 12122–12131. [CrossRef]
27. Burress, H.; Taylor, M.; Banerjee, T.; Tatulian, S.A.; Teter, K. Co- and Post-Translocation Roles for HSP90 in Cholera Intoxication. *J. Biol. Chem.* **2014**, *289*, 33644–33654. [CrossRef]
28. Kellner, A.; Taylor, M.; Banerjee, T.; Britt, C.B.T.; Teter, K. A Binding Motif for Hsp90 in the A Chains of ADP-Ribosylating Toxins That Move from the Endoplasmic Reticulum to the Cytosol. *Cell. Microbiol.* **2019**, *21*, e13074. [CrossRef]
29. Kellner, A.; Cherubin, P.; Harper, J.K.; Teter, K. Proline Isomerization as a Key Determinant for Hsp90-Toxin Interactions. *Front. Cell. Infect. Microbiol.* **2021**, *11*, 771653. [CrossRef]
30. Taylor, M.; Navarro-Garcia, F.; Huerta, J.; Burress, H.; Massey, S.; Ireton, K.; Teter, K. Hsp90 Is Required for Transfer of the Cholera Toxin A1 Subunit from the Endoplasmic Reticulum to the Cytosol. *J. Biol. Chem.* **2010**, *285*, 31261–31267. [CrossRef]
31. Dmochewitz, L.; Lillich, M.; Kaiser, E.; Jennings, L.D.; Lang, A.E.; Buchner, J.; Fischer, G.; Aktories, K.; Collier, R.J.; Barth, H. Role of CypA and Hsp90 in Membrane Translocation Mediated by Anthrax Protective Antigen. *Cell. Microbiol.* **2011**, *13*, 359–373. [CrossRef] [PubMed]
32. Steinemann, M.; Schlosser, A.; Jank, T.; Aktories, K. The Chaperonin TRiC/CCT Is Essential for the Action of Bacterial Glycosylating Protein Toxins like Clostridium Difficile Toxins A and B. *Proc. Natl. Acad. Sci. USA* **2018**, *115*, 9580–9585. [CrossRef] [PubMed]

33. Williamson, D.S.; Borgognoni, J.; Clay, A.; Daniels, Z.; Dokurno, P.; Drysdale, M.J.; Foloppe, N.; Francis, G.L.; Graham, C.J.; Howes, R.; et al. Novel Adenosine-Derived Inhibitors of 70 KDa Heat Shock Protein, Discovered through Structure-Based Design. *J. Med. Chem.* **2009**, *52*, 1510–1513. [CrossRef] [PubMed]
34. Rosenzweig, R.; Nillegoda, N.B.; Mayer, M.P.; Bukau, B. The Hsp70 Chaperone Network. *Nat. Rev. Mol. Cell. Biol.* **2019**, *20*, 665–680. [CrossRef] [PubMed]
35. Schuster, M.; Schnell, L.; Feigl, P.; Birkhofer, C.; Mohr, K.; Roeder, M.; Carle, S.; Langer, S.; Tippel, F.; Buchner, J.; et al. The Hsp90 Machinery Facilitates the Transport of Diphtheria Toxin into Human Cells. *Sci. Rep.* **2017**, *7*, 613. [CrossRef] [PubMed]
36. Concilli, M.; Petruzzelli, R.; Parisi, S.; Catalano, F.; Sirci, F.; Napolitano, F.; Renda, M.; Galietta, L.J.V.; Di Bernardo, D.; Polishchuk, R.S. Pharmacoproteomics Pinpoints HSP70 Interaction for Correction of the Most Frequent Wilson Disease-Causing Mutant of ATP7B. *Proc. Natl. Acad. Sci. USA* **2020**, *117*, 32453–32463. [CrossRef]
37. Clerico, E.M.; Tilitsky, J.M.; Meng, W.; Gierasch, L.M. How Hsp70 Molecular Machines Interact with Their Substrates to Mediate Diverse Physiological Functions. *J. Mol. Biol.* **2015**, *427*, 1575–1588. [CrossRef]
38. Finka, A.; Sharma, S.K.; Goloubinoff, P. Multi-Layered Molecular Mechanisms of Polypeptide Holding, Unfolding and Disaggregation by HSP70/HSP110 Chaperones. *Front. Mol. Biosci.* **2015**, *2*, 29. [CrossRef]
39. Balchin, D.; Hayer-Hartl, M.; Hartl, F.U. In Vivo Aspects of Protein Folding and Quality Control. *Science* **2016**, *353*, aac4354. [CrossRef]
40. Yam, A.Y.; Xia, Y.; Lin, H.-T.J.; Burlingame, A.; Gerstein, M.; Frydman, J. Defining the TRiC/CCT Interactome Links Chaperonin Function to Stabilization of Newly Made Proteins with Complex Topologies. *Nat. Struct. Mol. Biol.* **2008**, *15*, 1255–1262. [CrossRef]
41. Howe, M.K.; Bodoor, K.; Carlson, D.A.; Hughes, P.F.; Alwarawrah, Y.; Loiselle, D.R.; Jaeger, A.M.; Darr, D.B.; Jordan, J.L.; Hunter, L.M.; et al. Identification of an Allosteric Small-Molecule Inhibitor Selective for the Inducible Form of Heat Shock Protein 70. *Chem. Biol.* **2014**, *21*, 1648–1659. [CrossRef] [PubMed]
42. Reddymasu, S.C.; Soykan, I.; McCallum, R.W. Domperidone: Review of Pharmacology and Clinical Applications in Gastroenterology. *Am. J. Gastroenterol.* **2007**, *102*, 2036–2045. [CrossRef] [PubMed]
43. Ou, L.B.; Moriello, C.; Douros, A.; Filion, K.B. Domperidone and the Risks of Sudden Cardiac Death and Ventricular Arrhythmia: A Systematic Review and Meta-Analysis of Observational Studies. *Br. J. Clin. Pharm.* **2021**, *87*, 3649–3658. [CrossRef] [PubMed]
44. Johnson, S.; Gerding, D.N. Bezlotoxumab. *Clin. Infect. Dis.* **2019**, *68*, 699–704. [CrossRef] [PubMed]
45. Ernst, K.; Sailer, J.; Braune, M.; Barth, H. Intoxication of Mammalian Cells with Binary Clostridial Enterotoxins Is Inhibited by the Combination of Pharmacological Chaperone Inhibitors. *Naunyn Schmiedebergs Arch. Pharm.* **2021**, *394*, 941–954. [CrossRef]
46. Ernst, K.; Kling, C.; Landenberger, M.; Barth, H. Combined Pharmacological Inhibition of Cyclophilins, FK506-Binding Proteins, Hsp90, and Hsp70 Protects Cells from Clostridium Botulinum C2 Toxin. *Front. Pharmacol.* **2018**, *9*, 1287. [CrossRef] [PubMed]
47. Papatheodorou, P.; Zamboglou, C.; Genisyuerek, S.; Guttenberg, G.; Aktories, K. Clostridial Glucosylating Toxins Enter Cells via Clathrin-Mediated Endocytosis. *PLoS ONE* **2010**, *5*, e10673. [CrossRef]
48. von Eichel-Streiber, C.; Harperath, U.; Bosse, D.; Hadding, U. Purification of Two High Molecular Weight Toxins of Clostridium Difficile Which Are Antigenically Related. *Microb. Pathog.* **1987**, *2*, 307–318. [CrossRef]
49. Heber, S.; Barthold, L.; Baier, J.; Papatheodorou, P.; Fois, G.; Frick, M.; Barth, H.; Fischer, S. Inhibition of *Clostridioides difficile* Toxins TcdA and TcdB by Ambroxol. *Front. Pharmacol.* **2022**, *12*, 809595. [CrossRef]
50. Jank, T.; Ziegler, M.O.P.; Schulz, G.E.; Aktories, K. Inhibition of the Glucosyltransferase Activity of Clostridial Rho/Ras-Glucosylating Toxins by Castanospermine. *FEBS Lett.* **2008**, *582*, 2277–2282. [CrossRef]
51. Barth, H.; Blocker, D.; Behlke, J.; Bergsma-Schutter, W.; Brisson, A.; Benz, R.; Aktories, K. Cellular Uptake of Clostridium Botulinum C2 Toxin Requires Oligomerization and Acidification. *J. Biol. Chem.* **2000**, *275*, 18704–18711. [CrossRef] [PubMed]

toxins

MDPI

*Article*

# *Clostridium perfringens* Epsilon Toxin Binds to and Kills Primary Human Lymphocytes

**Samantha V. Shetty [1], Michael R. Mazzucco [1], Paige Winokur [2], Sylvia V. Haigh [1], Kareem Rashid Rumah [3], Vincent A. Fischetti [3], Timothy Vartanian [1] and Jennifer R. Linden [1,*]**

1 Feil Family Brain and Mind Research Institute, Weill Cornell Medical College, New York, NY 10065, USA; tiv2002@med.cornell.edu (T.V.)
2 Harold and Margaret Milliken Hatch Laboratory of Neuro-Endocrinology Rockefeller University, New York, NY 10065, USA
3 Laboratory of Bacterial Pathogenesis and Immunology, Rockefeller University, New York, NY 10065, USA
* Correspondence: jel2049@med.cornell.edu; Tel.: +1-206-714-0908

**Abstract:** *Clostridium perfringens* epsilon toxin (ETX) is the third most lethal bacterial toxin and has been suggested to be an environmental trigger of multiple sclerosis, an immune-mediated disease of the human central nervous system. However, ETX cytotoxicity on primary human cells has not been investigated. In this article, we demonstrate that ETX preferentially binds to and kills human lymphocytes expressing increased levels of the myelin and lymphocyte protein MAL. Using flow cytometry, ETX binding was determined to be time and dose dependent and was highest for CD4+ cells, followed by CD8+ and then CD19+ cells. Similar results were seen with ETX-induced cytotoxicity. To determine if ETX preference for CD4+ cells was related to MAL expression, MAL gene expression was determined by RT-qPCR. CD4+ cells had the highest amount of Mal gene expression followed by CD8+ and CD19+ cells. These data indicate that primary human cells are susceptible to ETX and support the hypothesis that MAL is a main receptor for ETX. Interestingly, ETX bindings to human lymphocytes suggest that ETX may influence immune response in multiple sclerosis.

**Keywords:** epsilon toxin; ETX; myelin and lymphocyte protein; MAL; clostridium perfringens; lymphocytes; CD4+ cells; CD8+ cells; CD19+ cells; B cells; T cells

**Key Contribution:** For the first time, we demonstrate that human primary cells are susceptible to ETX binding and ETX-induced cell death and positively associate with MAL gene expression in CD4+, CD8+, and CD19+ lymphocytes. These data support the hypothesis that MAL is a main receptor for ETX.

## 1. Introduction

Epsilon toxin (ETX) is produced by *Clostridium perfringens* (*C. perfringens*) toxinotypes B and D and is the third most lethal bacterial toxin known to date. Recently, ETX has been identified as a possible environmental cause of multiple sclerosis (MS) in humans, an immune-mediated, demyelinating disease of the central nervous system (CNS) [1–3]. Our group has demonstrated that MS patients are more likely to be colonized by ETX-producing *C. perfringens* toxinotypes versus healthy controls (HC), 61% versus 13%, respectively [3]. In addition, in colonized individuals, the relative abundance of ETX producing strains is significantly higher in MS patients versus HC. Histopathological, active MS lesions are characterized by overt blood–brain-barrier permeability, demyelination, and robust CNS immune infiltration. Using both in vitro and in vivo models, we have demonstrated that ETX specifically causes BBB permeability [4,5], demyelination [6], and loss of CNS-immune privilege [3]. Remarkable, this specificity is mediated by the selective expression of the ETX receptor, the myelin and lymphocyte protein MAL [7], on CNS endothelial cells [5], mature oligodendrocytes [6], and human lymphocytes [8,9] (murine lymphocytes do not express MAL [10]). Importantly, pathogenic lymphocytes, including both T cells and B cells, play an important role in MS pathogenesis [11,12]. However, whether ETX targets primary human lymphocytes or plays a role in MS pathogenesis is currently unknown. In this paper, we sought to determine if ETX can bind to and kill primary human lymphocytes expressing MAL.

ETX is secreted as a relatively inactive 33 kDa protoxin (pETX), and proteolytic cleavage at the N and C terminal by trypsin and chymotrypsin (or bacterial lambda toxin) results in 1000 fold increase in ETX's toxicity [13]. Cytotoxicity is mediated by pore formation/oligomerization, resulting in ionic imbalances [14–16]. In ruminant livestock, ETX is responsible for causing an economically devastating disease, enterotoxaemia [17,18]. In both natural and experimental infections, ETX preferentially accumulates in the brains and kidneys of animals [19–25]. In the brain, ETX selectively binds to brain endothelial cells, myelinated structures, and mature oligodendrocytes, the myelinating cells of the CNS [5,6,26–30]. In the kidney, ETX preferentially binds to distal and collecting tubules [22,24]. ETX preferences for these specific cell types are most likely due to the expression of the ETX receptor MAL.

Although a handful of proteins have been identified as possible receptors of co-receptors for ETX [7,31,32], MAL expression is the only protein to have been demonstrated to be both necessary and sufficient for ETX binding and cell death. Importantly, this protein has been shown to be necessary for ETX binding and cytotoxicity in both gain-of-function and loss-of-function assays and has been demonstrated to directly interact with ETX via immunoprecipitation assays. Expression of both human and rodent MAL in the normally ETX-resistant Chinese hamster (CHO) line conferred both ETX binding and cytotoxicity [7]. In addition, MAL deficiency (knockout) abolished ETX binding to tissue [5,7]. Recently, these results have been confirmed in an independent study that again demonstrated that (1) the expression of endogenous human MAL in ETX-resistant cell lines confirmed ETX binding and cytotoxicity, and (2) the deletion of MAL expression in normally ETX-sensitive cell lines abolished ETX binding and cytotoxicity [9]. Importantly, these authors were able to demonstrate the direct protein–protein interaction of MAL and ETX via immunoprecipitation. Finally, MAL expression was also necessary for ETX-induced oligodendrocyte cell death and demyelination as well as the ETX-induced blood–brain barrier permeability [5,6]. Based on these various loss-of-function and gain-of-function assays, we believe that MAL is the strongest candidate as the main ETX receptor.

MAL is a proteolipid, tetraspan protein with two extracellular loops. MAL cDNA was first cloned from human T cell lines and is indicated to play an important role in T cell differentiation, membrane protein sorting, and formation of immunological synapses [8,33–38]. Since then, MAL has also been cloned from oligodendrocytes, myelin, MDCK cells, whole mouse brains, and found to be expressed in certain polarized epithelial cells [8,39–43]. MAL localizes to lipid rafts and plays an important role in lipid raft formation, stabilization, and maintenance [37,44–46]. In addition, MAL plays a significant role in protein cycling from the trans-golgi network to the plasma membrane and is important for apical protein sorting in polarized cells [42,46–53]. In the nervous system, MAL is believed to play an important role in myelin biogenesis and proper axon–glial interactions [41,54].

In human lymphocytes, MAL expression has been reported to be confined to the T cell lineage; expression of MAL in healthy B cell lineages is less clear [8,33–35,37,38,55–59]. Recently, it has been demonstrated that ETX can bind to and kill cancerous cell lines of the T cell lineage which express MAL but not cell lines of the B cell lineage which do not express MAL [9]. However, as MAL expression has been reported to be dysregulated in cancer, including cancerous lymphocytic cell lines [56,60–65], it is imperative that ETX-lymphocyte interactions be confirmed in primary, human lymphocytes isolated from healthy donors. The aim of this study was to determine if ETX could bind to and kill primary human lymphocytes that endogenously express MAL, specifically, human CD4+, CD8+, and CD19+ lymphocytes. We hypothesize that lymphocytes with higher amounts of MAL expression will have a higher affinity for and thus an increased susceptibility to ETX.

## 2. Results

### 2.1. Primary Human Lymphocytes Express Mal

To confirm *Mal* gene expression in the T cell lineage, real-time quantitative PCR (RT-qPCR) for human *Mal* was performed on isolated CD4+, CD8+, and B cells (Figure 1). *Mal* gene expression was normalized to CD4+ cells. RT-qPCR analysis confirmed that CD4+ cells had the highest amount of *Mal* gene expression compared to isolated CD8+ and B cells (Figure 1). In addition, CD8+ cells showed a trend towards expressing significantly more *Mal* than B cells.

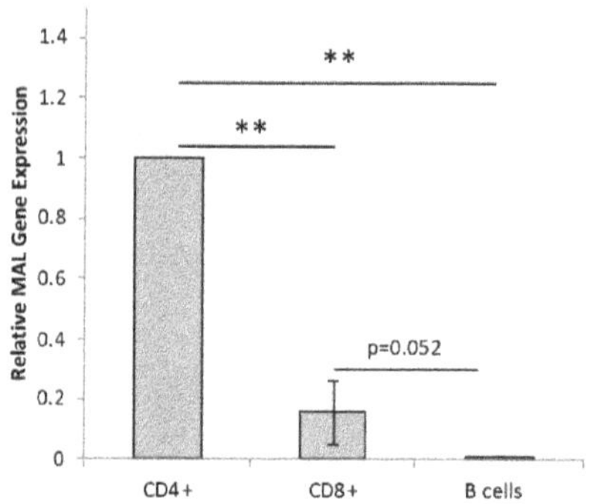


**Figure 1.** Real-time quantitative PCR (RT-qPCR) analysis of *Mal* gene expression in primary human lymphocytes. *Mal* transcripts levels were quantified in cDNAs obtained from isolated populations of CD4+, CD8+, and B cells. Relative *Mal* expression in isolated CD4+, CD8+, and B cells. B-actin was used as a reference gene. ** $p < 0.01$ determined by One-way ANOVA with post hoc Tukey HSD Test. Results are expressed as the mean performed in triplicate.

The detection of *Mal* expression in B cells was unexpected and may be due to T cell contamination during our isolation method. To determine if the low Mal expression observed in our isolated B cells populations were a result of T cell contamination, we compared our *Mal* expression results to those of other publicly available datasets using a variety of cell isolation and gene expression techniques (Supplemental Figure S1) [66–69]. Isolation methods included FACS sorting (Supplemental Figure S1A,B), positive magnetic selection (Supplemental Figure S1C), and single-cell RNAseq analysis (Supplemental Figure S1D–F), whereas *Mal* expression was evaluated using RNAseq (Supplemental Figure S1A,B,D) and a microarray (Supplemental Figure S1C). The examination of these four independent datasets confirmed significantly higher *Mal* gene expression in CD4+ cells, followed by CD8+ cells, and, finally, CD19+/B cells. These results also demonstrated a low but still detectable level of *Mal* transcripts in CD19/B cells, consistent with our RT-qPCR results. Based on these findings, we believe our RT-qPCR results are accurate.

### 2.2. ETX Binds to Human Lymphocytes with a Preference for CD4+ Cells

To determine if the ETX bound to human lymphocytes expressing *Mal,* PBMNCs were probed with 50 nM of Alexa Fluor 647 pETX (pETX-647) for 2 h, and binding to CD4+, CD8+, and CD19+ cells evaluated by multicolor flow cytometry (Figure 2A–C). pETX was used to study ETX binding because pETX binds with a similar affinity as active ETX but does not oligomerize and form pores, preventing endosome recycling and the possible cell surface rearrangement of MAL [5,7,15,34,37,38,70–77]. Untreated cells (0 nM) were used as negative controls. Scatter plots (Figure 2A) and histogram analyses of pETX-647 fluorescent intensities (Figure 2B) revealed that CD4+ cells bound more toxins than CD8+ and CD19+ cells. In addition, CD8+ cells bound more toxins than CD19+ cells.

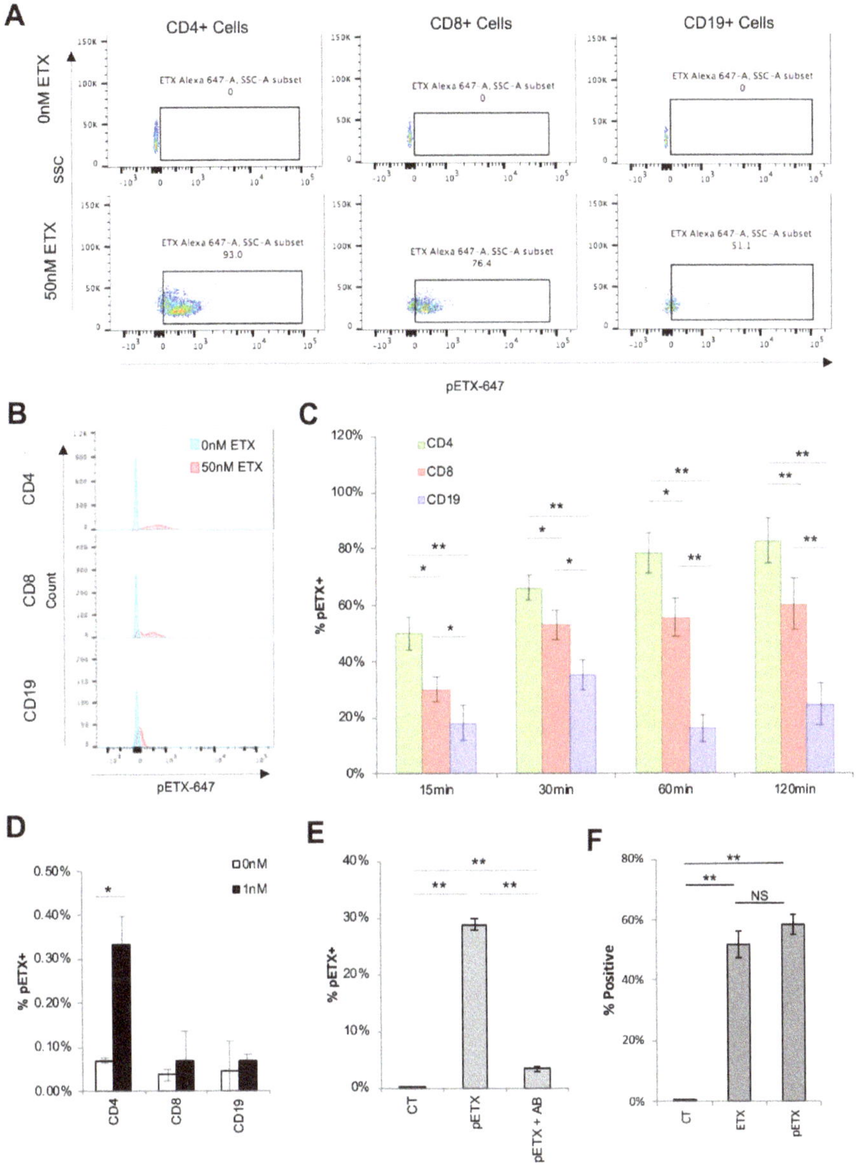


**Figure 2.** ETX binds to CD4+, CD8+, and CD19+ lymphocytes with a preference for CD4+ cells. PBMNCs were incubated with 0 nM or 50 nM pETX-647 for 2 h and binding to CD4+, CD8+, and CD19+ cells was examined by flow cytometry. An example of the gating strategy for examination of CD4+, CD8+, and CD19+ lymphocytes is depicted in Supplementary Figure S2. Representative scatter plots (**A**) and histogram analysis (**B**) of pETX-647 fluorescent intensity from three separate donors performed in triplicate. (**C**) PBMNCs were incubated with 25 nM pETX-647 for indicated time points and binding determined by flow cytometry. Results are expressed as the percent of CD4+, CD8+, or CD19+ cells positive for pETX (% pETX+). Results are the means of three separate donors performed in triplicate. (**D**) PBMNCs were incubated with 0 nM or 1 nM pETX-647 for 2 h. Results

are expressed as the percentages of CD4+, CD8+, or CD19+ cells positive for pETX (% pETX+). Results are the means of two separate donors performed in triplicate. (**E**) Anti-ETX antibody inhibits binding to CD4+ cells. Media containing 50 nM pETX-647 was pretreated with or without an antibody known to block ETX binding for 30 min before treating PBMNC for 2 h and evaluated by flow cytometry. Percent of pETX+ lymphocytes when cells are incubated without pETX-647 (CT), with pETX-647 (pETX), or pETX-647 pretreated with anti-ETX antibody (pETX + anti-ETX). (**F**) PBMNCs were treated with 25 nM of unlabeled pETX or ETX for 2 h. Untreated cells were used as controls. pETX and ETX binding to lymphocytes was determined using an affinity-purified anti-ETX polyclonal rabbit antibody and PE-conjugated anti-rabbit IgG. Results expressed as percent CD4+ lymphocytes positive for ETX or pETX (% Positive). Data points are the mean performed in triplicate. * $p < 0.05$, ** $p < 0.01$, determine by One-way ANOVA with post hoc Tukey HSD Test.

To quantify ETX bindings to target cells, pETX-647 binding was quantified via flow cytometry after 2 h of incubation with 25 nM of pETX-647. Significantly more CD4+ cells were positive for pETX-647 compared to CD8+ and CD19+ cells; 82.3%, 60.3%, and 24.7%, respectively (Figure 2C). Even when cells were incubated with 25 nM of pETX-647 for 15 min, 50.1%, 30.0%, and 18.2% of CD4+, CD8+, and CD19+ cells were positive for pETX-647, respectively. This trend was observed for all investigated time points.

Remarkably, pETX-647 could be observed binding to CD4+ cells at concentrations as low as 1 nM (Figure 2D). PBMNCs were incubated with 1 nM of pETX-647 for 2 h, and percent positive cells were evaluated by flow cytometry. Untreated cells (0 nM) were used as negative controls. No significant differences were observed in CD8+ or CD19+ when cells were treated with or without 1 nM pETX-647. In comparison, significantly more CD4+ cells were positive for pETX-647 when treated with 1 nM pETX-647 than without, 0.34% versus 0.07%, respectively.

To confirm that pETX bindings to lymphocytes is pETX-specific, pETX-647 was pretreated with an anti-ETX antibody shown to block ETX binding [78]. Binding to total lymphocytes was inhibited when media containing pETX-647 was pre-treated with the anti-ETX antibody (Figure 2E). To ensure that the fluorescent signal observed in lymphocytes was not due to excess fluorophore from the pETX-647 labeling process, PBMNCs were treated with shiga toxin (STX) fluorescently conjugated with Alexa Fluor 647 using the exact same labeling process (STX-647) (Supplemental Figure S3). When PBMNCs were incubated with 50 nM of STX-647 for 2 h, only a small percentage of lymphocytes bound STX-647, less than 2%. Importantly, the percentage of CD19+ cells positive for STX-647 was significantly higher than CD4+ or CD8+ cells: 1.64%, 0.19%, and 0.01%, respectively, confirming previous results that B cells had increased affinity for STX [79]. The low percentage of STX-647-positive cells indicates that the contamination of lymphocytes by excess dye when treated with Alexa Fluor conjugated 647 toxin is minimal.

Finally, we sought to confirm that probing cells with pETX was a reliable marker for ETX binding. PBMNCs were probed with 25 nM ETX or pETX for 2 h, and bindings were determined using an affinity purified anti-ETX polyclonal rabbit antibody and PE-conjugated anti-rabbit IgG and examined by flow cytometry (Figure 2F). Cells treated without ETX were used as controls. No significant differences in ETX and pETX binding were observed. Similar results were obtained when cells were probed with 5 nM, 10 nM, and 50 nM of ETX or pETX as well (data not shown). This confirms the previously published results that pETX and ETX bound similarly to target cells [4,24].

Taken together, this data indicates that ETX specifically binds to human primary lymphocytes, with a preference for CD4+ cells, followed by CD8+ and then CD19+ cells. Importantly, ETX binding positively associates with *Mal* gene expression.

### 2.3. ETX Bindings to Human Lymphocytes Is Dose and Time Dependent

To determine if the ETX bindings to lymphocyte subsets were dose and time dependent, PBMNCs were incubated with 0 nM, 1 nM, 5 nM, 10 nM, 25 nM, and 50 nM of pETX-647 for 15, 30, 60, and 120 min. After 15 min of incubation, pETX was observed binding to CD4+, CD8+, and CD19+ cells in a dose-dependent manner (Figure 3A). For a full breakdown of *p* values between different doses, please refer to Supplementary Table S1. After 15 min, significantly more CD4+ cells were positive for ETX when treated with 10 nM (11%), 25 nM (50%), and 50 nM (75%) compared to untreated controls (0%) (Figure 3A). In comparison, a significant increase in ETX-positive CD8+ and CD19+ cells was not observed until treatment with 25 nM pETX-647. When PBMNCs were incubated for 120 min with pETX-647, significantly more CD4+ cells were positive for ETX at 5 nM (41%), 10 nM (62%), 25 nM (83%), and 50 nM (90%) compared to untreated controls (0%) (Figure 3B). In contrast, a significant increase in pETX-647 positive CD8+ and CD19+ cells was not observed until cells were treated with 10 nM pETX-647. A similar trend was observed when cells were incubated with pETX-647 for 30 and 60 min (Supplemental Figure S4a,b, respectively). These data indicated that the ETX bindings to all lymphocyte subsets is dose dependent and reaffirms that ETX preferentially binds to CD4+ compared to CD8+ and CD19+ cells.

For CD4+ cells, significant differences in ETX bindings at different time points were observed when cells were treated with 1 nM, 5 nM, 10 nM, or 25 nM pETX-647 (Figure 3C). Remarkably, after 1 nM treatment for 15 min, 0.021% of CD4+ cells were positive for ETX. After 60 and 120 min, significantly more CD4+ cells were positive for pETX: 0.060% and 0.067%, respectively. In comparison, with 5 nM treatment, 3.2%, 8.1%, 23.8%, and 40.9% of CD4+ cells were positive for pETX after 15-, 30-, 60-, and 120-min incubations, respectively. Similar trends were seen for 10 nM and 25 nM doses. At 50 nM pETX-647 treatment, pETX binding to CD4+ cells appeared to be saturated, as there were no significant differences between any of the time points. Similar results were observed in CD8+ cells (Figure 3D), with the clearest time-dependent binding occurring with 10 nM pETX-647 treatment. In total, 4.4%, 17.2%, 21.6%, and 34.2% of CD8+ cells were positive for pETX after 15-, 30-, 60-, and 120-min incubations, respectively. Again, pETX binding appeared to be saturated for all time points at 50 nM treatment for CD8+ cells. Although ETX binding was observed on CD19+ cells, binding did not appear to be time dependent under these conditions (Figure 3E). These data indicated that the ETX bindings to CD4+ and CD8+ cells were time dependent and, again, reaffirmed that ETX preferentially bound to CD4+ cells, compared to CD8+ and CD19+ cells.

### 2.4. ETX Induces Cytotoxicity in Human Lymphocytes, Especially CD4+ Cells

To determine if ETX binding to human lymphocytes conferred cytotoxicity, total lymphocytes were evaluated for cell death by propidium iodide (PI) inclusion via flow cytometry (Figure 4A,B). Cells positive for (PI+) were considered dead. After 4 h of active ETX treatment, only a small percentage of cell death was observed: 1.4% (Figure 4C). A significant increase in total lymphocyte cell death was observed at ETX doses of 25 nM and 50 nM: 11.5% and 17.8%, respectively. Importantly, pretreatment of ETX with a neutralizing antibody that blocks ETX-cytotoxicity inhibited ETX-induced cell death (Figure 4D) [78].

To determine if lymphocyte populations expressing higher levels of *Mal* were more susceptible to ETX-induced cytotoxicity, cell death was evaluated in CD4+, CD8+ and CD19+ cells by flow cytometry (Figure 4E). When treated with 25 nM of ETX for 4 h, cell death was significantly higher in CD4+ cells compared to CD8+ and CD19+ cells: 19.6%, 6.5%, and 3.0%, respectively. Similar results were seen when cells were treated with 50 nM ETX with CD4+, CD8+, and CD19+ cells, exhibiting 35.1%, 13.9%, and 1.9% cell death, respectively. In addition, CD8+ cell death was significantly higher than CD19+ cell death. Taken together, these data demonstrate that ETX induced cell death in CD4+ and CD8+ cells is positively associated with *Mal* gene expression.

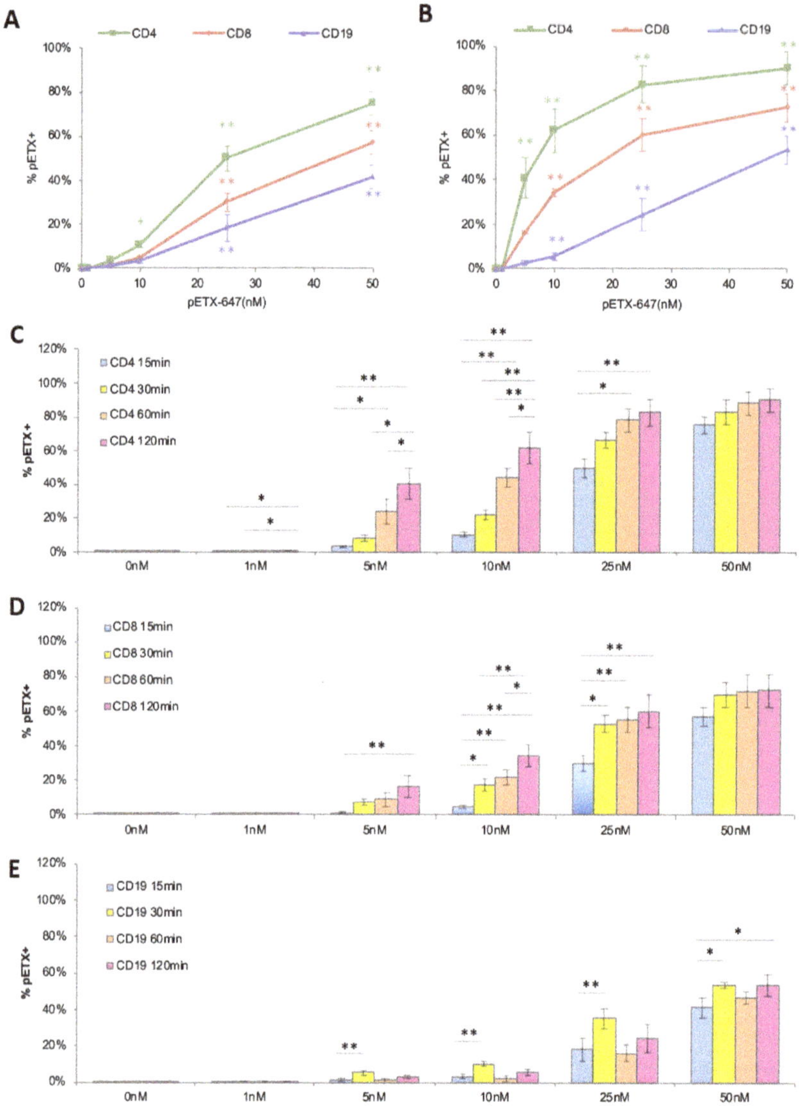


**Figure 3.** ETX bindings to lymphocytes are time and dose dependent. To determine if ETX binding is dose dependent, PBMNCs were incubated with indicated doses of pETX-647 for 15 (**A**) and 120 min (**B**), and pETX-647 binding was determined by flow cytometry. Results are expressed as percent pETX positive (% pETX+) cells for CD4+, CD8+, and CD19+ cells. * $p < 0.05$ and ** $p < 0.001$ compared to untreated controls (0 nM), as determined by One-way ANOVA with post hoc Tukey HSD Test. For a more detailed analysis of $p$ values for all pETX doses, refer to supplementary Table S1. To determine if ETX binding is time dependent, PBMNCs were incubated with 0 nM, 1 nM, 5 nM, 10 nM, 25 nM, or 50 nM pETX-647 for indicated time points (**C**–**E**). The percentages of ETX positive cells were determined by flow cytometry for CD4+ cells (**C**), CD8+ cells (**D**), and CD19+ cells (**E**). * $p < 0.01$ and ** $p < 0.001$ were determined by One-way ANOVA with post hoc Tukey HSD Test. All results are the mean results of three donors performed in triplicate.

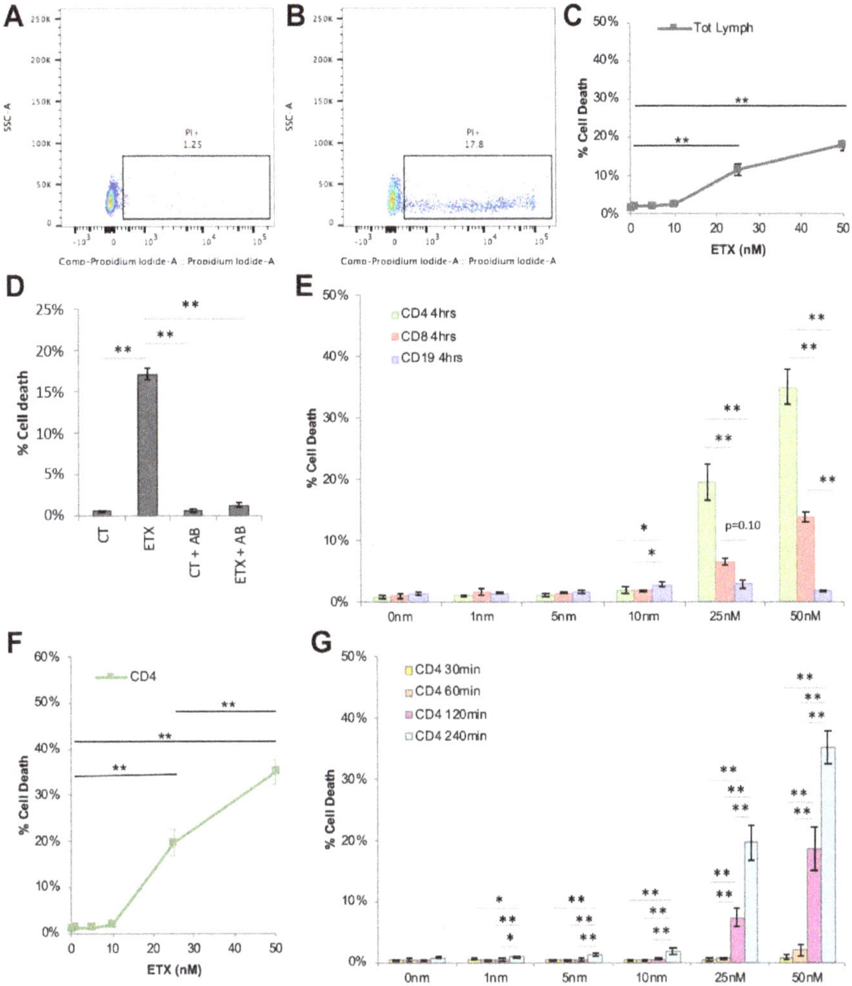

**Figure 4.** Active ETX kills human CD4+ cells in a dose- and time-dependent manner. PI inclusion was used to evaluate cell viability via flow cytometry. PI positive cells are considered dead. Results are expressed as the number of PI positive cells out of the total specific population and expressed as percent cell death (% cell death). (**A**,**B**) Representative scatter plots of total lymphocytes stained with PI to evaluate cell death after treatment with 0 nM (**A**) or 50 nM (**B**) ETX for 4 h. (**C**) Total lymphocyte cell death after 4 h of incubation with indicated ETX doses. (**D**) Media containing 50 nM ETX were pretreated with or without a neutralizing anti-ETX antibody for 30 min before treating lymphocytes for 2 h. Cell death was evaluated by PI inclusion via flow cytometry. Percent cell death when lymphocytes are treated in media alone (CT), media containing 50 nM ETX (ETX), media pretreated with anti-ETX antibody (CT + anti-ETX), and media containing 50 nM ETX pretreated with anti-ETX antibody (ETX + anti-ETX). (**E**) Percent cell death of different lymphocyte subsets at indicated ETX doses after 4 h of treatment. (**F**) Cell death was evaluated in CD4+ cells after 4 h of treatment at indicated ETX doses. These data are the same data depicted in Figure 4E with different statistical analyses. ETX-induced cell death of CD4+ cells is dose dependent. (**G**) ETX-induced cytotoxicity in CD4+ cells over time at various time points. Data are the mean of a single donor in triplicate. Data are representative of multiple donors. * $p < 0.05$, ** $p < 0.01$, determined by One-way ANOVA with post hoc Tukey HSD Test.

### 2.5. ETX-Induced Cytotoxicity in Human CD4+ Cells Is Time and Dose Dependent

To determine if ETX-induced cytotoxicity in CD4+ cells is dose dependent, PBMNCs were incubated with 0 nM, 1 nM, 5 nM, 10 nM, 25 nM, and 50 nM of active ETX for 4 h (Figure 4F). Significant increases in percent cell death compared to untreated controls (0.82%) were observed at 25 nM (19.6%) and 50 nM (35.1%) doses. Cell death at 50 nM was significantly higher than at 25 nM treatment. This indicates that ETX-induced cell death of CD4+ cells is dose dependent.

To determine if ETX-induced cell death of CD4+ cells is time-dependent, PBMNCs were incubated with indicated doses of ETX for 30, 60, 120, and 240 min (Figure 4G). At a dose as low as 1 nM, a significant increase in CD4+ cell death was observed between 30 min and 4 h: 0.55% and 1%, respectively. At a dose of 5 nM, a significant increase in CD4+ cell death was observed between 30 min and 4 h: 0.47% and 1.21%, respectively. Cell death was considerably higher at larger ETX doses. At 50 nM ETX treatment, cell death was 0.84%, 2.06%, 18.06%, and 35.07% after 30 min, 60 min, 2 h, and 4 h of treatment, respectively. This data indicates that ETX induced CD4+ cell death in a time-dependent.

### 2.6. ETX-Induced Cytotoxicity in Human Lymphocytes Is Mediated by Pore Formation

ETX-induced cell death has been proposed to be mediated by pore formation in sensitive cell lines. To determine if ETX pore formation occurs in primary human lymphocytes, whole-cell lysates from PBMNC treated with 0 nM, 10 nM, 25 nM, or 50 nM ETX for two hours were evaluated by Western blot analysis (Figure 5A). Lysates from control and ETX-treated rMAL-CHO cells, known to form a 150 kDA ETX pore complex [78], were used as positive controls. The 150 kDA pore complex was only observed when PBMNCs were treated with 50 nM ETX (Figure 5A). In addition, a band at 27 kDA could be observed in all cell lysates treated with 50 and 25 nM ETX, indicating bound ETX monomers, as all cells were thoroughly washed in PBS prior to lysis. Pore formation appeared to be time dependent when cells were treated with 50 nM of ETX for 30, 60, and 120 min (Figure 5B). This indicates that ETX-mediated cytotoxicity of PBMNCs is mediated by pore formation.

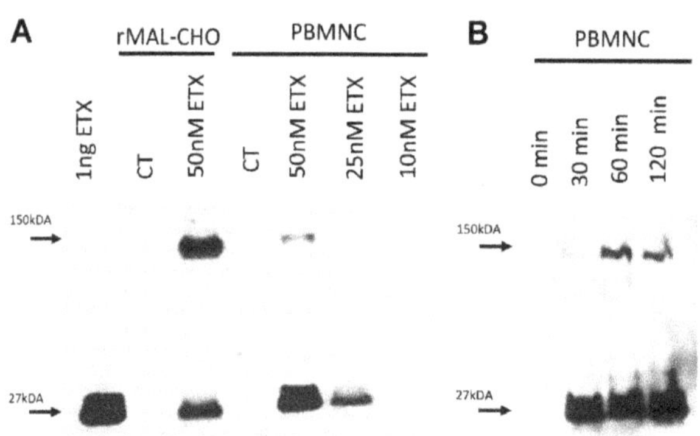


**Figure 5.** ETX cytotoxicity is mediated by pore formation in PBMNC. (**A**) PBMNCs were treated with indicated doses of ETX for 2 h. Cells were extensively washed in PBS, and whole-cell lysates were examined via Western blot for detection of the 150kDa oligomerized pore complex or 27 kDa bound ETX monomer. In total, 1 ng of ETX and whole-cell lysates from rMAL-CHO cells treated with or without 50 nM ETX were used as controls. (**B**) PBMNCs were treated with 50 nM of ETX for the indicated time points and examined for via Western blot for detection of the 150 kDa oligomerized pore complex or 27 kDa bound monomer.

## 3. Discussion

### 3.1. ETX Binding and Cytotoxicity Positively Associates with Mal Gene Expression in Human Lymphocytes

In this paper, we demonstrate that ETX binding and cytotoxicity to primary human lymphocyte populations positively associates with *Mal* gene expression. Specifically, CD4+ cells have the highest amount of *Mal* gene expression, followed by CD8+ and then B cells. Accordingly, ETX preferentially binds to and kills CD4+ cells, followed by CD8+ and then CD19+ cells. This builds on previously published data demonstrating that ETX can bind to and kill human cell lines of the cancerous T cell lineage and is MAL dependent [9].

We first confirmed *Mal* expression in CD4+, CD8+, and B cells using RT-qPCR. The results demonstrated that CD4+ had the highest amount of *Mal* gene expression, followed by CD8+ cells and then CD19+ cells. Increased *Mal* gene expression in CD4+ cells was also confirmed using publicly available datasets [66–69]. Secondly, we demonstrated that ETX bound to human lymphocytes with a preference that positively associates with *Mal* expression. When lymphocytes were incubated with 25 nM of pETX-647 for two hours, 82.8% of CD4+, 60.3% of CD8+, and 24.7% of CD19+ cells were bound to pETX. In addition, pETX binding occurred in a dose- and time-dependent manner. Finally, we demonstrated that ETX induced lymphocyte cell death. When lymphocytes were incubated with 50 nM of active ETX for four hours, 35.1% of CD4+ and 13.9% of CD8+ exhibited cell death; no significant amount of cell death was observed in CD19+ cells. ETX-induced cell death of CD4+ cells was also dose and time dependent. Taken together, this data indicates that ETX binding and cytotoxicity to human lymphocytes is positively associated with *Mal* expression. However, these results were limited, as they only analyzed *Mal* gene expression and not protein expression. Future experiments need to confirm that *Mal* gene expression is reflective of MAL protein expression, as we were unable to complete protein analysis experiments at this time. Alternatively, ETX's increased binding to and activity on CD4+ cells may have also been a result of CD4 expression itself. In a single experiment, ETX was observed to bind to recombinant human CD4 immobilized on Dynabeads [32], raising the possibility that CD4+ affinity and sensitivity to ETX may be a combination of both CD4 and MAL expression. Importantly, other ETX-sensitive cell lines, including MDCK and ACHN cells, also express MAL [42,47,80]. Taken together, these observations supported the theory that MAL is the main receptor for ETX.

### 3.2. MAL Expression in Human Lymphocytes

MAL expression in CD4+ and CD8+ cells was consistent with previously published results, looking at both peripheral blood lymphocytes and various cell lines of the T- and B-cell lineage. By using a privately generated anti-MAL antibody [81], Copie-Bergman et al. demonstrated that 65–90% of CD4+ and 22–39% of CD8+ cells were positive for MAL via flow cytometry [56]. In comparison, the authors did not detect a significant amount of MAL expression on B cells from peripheral blood, tonsils, or spleens: 0–0.6%, 1.5–2%, and 0.6–0.7%, respectively. They did, however, observe the occasional MAL-positive plasma cell via immunohistochemistry in tonsils or reactive lymph nodes. In addition, other groups have observed MAL expression in various T cell lines but not B cell lineages [8,33,35,37,57,58]. Conflicting results for MAL detection in B cells may have been a result of technical differences in experimental approaches and sensitivity (for example, protein expression versus gene expression). However, results consistently indicated that B cells expressed significantly lower to no MAL compared to T cells.

### 3.3. MAL's Function in Different Lymphocyte Populations

The reason for the differential expression of MAL in specific lymphocyte populations is unknown. MAL's function in lymphocytes has only been extensively studied in the T cell lineage. In general, MAL appears to play an important role in lipid raft formation and stabilization [36,37,44–46] and protein trafficking to the apical plasma membrane in polarized cells [42,46,48,49,51,52]. In human T cells, MAL is selectively present in glycolipid-enriched membrane microdomains (also known as detergent-resistant membranes) and appears to play an important role in T cell activation, mainly through its interactions with Lck, a src-like kinase [33–35,55].

Src-like kinases, especially Lck, play an essential role in T cell activation and maturation [82]. Previous studies have shown that MAL and Lck co-immunoprecipitate with each other in a lipid-dependent interaction in T cells [35]. If the expression of MAL is lost, Lck targeting to the plasma membrane becomes dysfunctional. As such, the loss of MAL results in the defective polarization of the T cell receptor for antigen (TCR) and the organization of the immunological synapse (IS) [33]. MAL targets Lck to the plasma membrane via vesicle movement along microtubule tracks and requires participation of Inverted Formin2 (INF2) as well as Cdc42 and Rac1 [55]. MAL has also been shown to be necessary for proper receptor and signaling protein assembly at the IS in the supramolecular activation cluster (SMAC). The incorrect localization of MAL results in Lck being transported to the wrong part of the SMAC [34]. In addition, more recent publications have also demonstrated that MAL plays an important role in endosome trafficking and exosome secretion from T cells [38].

Although MAL's function in B cells is unknown, it is possible that MAL could play a similar role in lipid raft-protein organization and signaling in B cells. Lipid rafts play a role in B cell activation and can act as platforms for B cell receptor (BCR) signaling and possibly antigen trafficking [83]. MAL may play a similar role in Lck or other src-like kinase trafficking in B cells. Interestingly, MAL is highly expressed in mediastinal large B cell lymphoma [59,64,84] and a subset of Hodgkin lymphoma with poor prognosis [65].

### 3.4. ETX-Induced Cell Death Pathways

It is generally accepted that ETX causes cell death via the formation/oligomerization of a heptameric pore. ETX pore formation occurs in three sequential steps: (1) ETX binding to its receptor, (2) oligomerization of the pre-pore complex on the cell surface, and (3) the pore insertion into the cell membrane [15]. Pore formation results in a rapid decrease in transmembrane resistance and the rapid depletion of intracellular $K^+$ and $Cl^-$ [85,86]. This is followed by a slower intracellular increase in $Na^+$ and $Ca^{2+}$ [85,87]. ETX also causes a rapid depletion of ATP and causes mitochondrial membrane permeabilization and translocation of an apoptotic-inducing factor to the nucleus [86].

In our current experiments, the ETX treatment of PBMNCs resulted in ETX oligomerization/pore formation, as detected by Western blot. However, the majority of the ETX detected in the PBMNC lysates was observed as bound monomers, not in the pore complex. It is interesting to note that 90% of CD4+ cells were positive when PBMNCs were probed with 50 nM of pETX-647 for 2 h. However, when cells were treated with 50 nM of active ETX for 4 h, only 35% of the CD4+ cells died, indicating a large discrepancy in ETX binding versus cell death in CD4+ cells for this dose and time point. Alternatively, in rMAL-CHO cells, a highly ETX-susceptible cell line, we saw a closer correlation of ETX binding and ETX cytotoxicity [7,78]. When rMAL-CHO cells were treated with ETX, the vast majority of cells bound to ETX and also died. For example, when treated with 50 nM of ETX, cell viability decreased to almost 0%. In addition, pore formation happened rapidly (within 5 min, when cells were treated with 50 nM ETX) and at very low doses (within 30 min after cells were treated with 1 nM ETX), with the majority of ETX detected in the poor complex, not as bound monomers [78]. This indicates that the low amount of cell death observed in CD4+ cells despite the high binding percentage may be due to the low amount of ETX oligomerization/pore formation observed in these cells.

### 3.5. Possible Role of ETX-Lymphocytes Interactions in MS Pathogenesis

Based on the limited amount of ETX-induced cell-death observed in human lymphocytes, despite a significantly higher degree of ETX binding, we postulate that ETX binding to lymphocytes may influence other cellular behaviors in addition to cell death, including various immune functions. This raises the interesting hypothesis that ETX bindings to MAL on human lymphocytes at sublethal doses may modify immune function, possibly influencing lymphocyte activity in immune-mediated diseases such as MS. CD4+, CD8+, and B cells have all been implicated in MS pathogenesis, although the exact mechanisms by which they influence MS pathogenesis is still unclear [11,12,88–93]. It is believed that pathogenic lymphocytes, including autoreactive and proinflammatory lymphocytes, are stimulated in the periphery, prior to the infiltration of these cells into the CNS. Histopathological examination of active MS lesions reveals dense lymphocytic infiltration into the CNS perivascular space with more limited extravasation into the CNS parenchyma. These infiltrates are heavily dominated by the presence of CD4+ and CD8+ with a much lower presence of B cells. Based on these observations, MS pathology has historically been viewed as being T cell driven; however, the wide success of B cell depleting therapies in treating MS has highlighted the importance of B cells in MS pathogenesis as well. Due to MAL's role in T cell activation and the important role it plays at the immunological synapse [33–35,55], it seems possible that ETX bindings to MAL may initiate a wide array of signaling cascades. Indeed, other pore-forming toxins have been shown to induce numerous cell signaling cascades not related to membrane permeabilization [94–103]. However, more examination into ETX's impact on lymphocyte function is needed and is an area of ongoing research.

### 3.6. Conclusions

In conclusion, we have demonstrated that ETX binding and cytotoxicity to human lymphocytes positively associates with MAL gene expression, supporting the hypothesis that MAL is the main receptor for ETX.

## 4. Materials and Methods

### 4.1. Peripheral Blood Isolation from Healthy Controls

Peripheral blood samples were collected from healthy controls via the cubital vein using BD Vacutainer K2 EDTA 7.2 mg Blood Collection tubes in accordance with our Institutional Review Board, protocol number 1003010940. At the time of donation, healthy controls were free of any chronic or acute disease, were both male and female, ranged in age from 18 to 59 years old, and lived in the New York City metropolitan area.

### 4.2. Isolation of CD4+, CD8+, and B Cells from Human Peripheral Blood for RT-qPCR Analysis

Peripheral blood samples were collected from healthy controls via the cubital vein using BD Vacutainer K2 EDTA 7.2 mg Blood Collection tubes. Subsets were isolated using RossetteSep immunodensity negative selection cocktails (Stem Cell Technologies, Vancouver, BC, Canada). CD4+, CD8+, and B cells were isolated using RosetteSep™ Human CD4+ T Cell Enrichment Cocktail, RosetteSep™ Human CD8+ T Cell Enrichment Cocktail, and RosetteSep™ Human B Cell Enrichment Cocktail, respectively, per manufacturer's instructions.

### 4.3. Real-Time Quantitative PCR (RT-qPCR) Analysis

Comparative RT-qPCR was performed in an ABI Taqman 7900HT Fast Real-Time PCR machine (Applied Biosystems, Waltham, MA, USA), using the PowerUp™ SYBR™ Green Master Mix (Applied Biosystems, Waltham, MA, USA). Briefly, PCR was carried out in a 10 μL volume in a final concentration of 1X SYBR™ Green Master Mix containing 300 nM forward and reverse primers and 10 ng cDNA. The primer sequences were as follows: human MAL, F 5′-GGGTGATGTTCGTGTCTGTG-3′, R 5′-ACTGAGGCGCTGAGGTAAAA-3′; human b-actin, F 5′-CACCAACTGGGACGACAT-3′, R 5′-ACAGCCTGGATAGCAACG-3′. The PCR reaction steps were as follows: 50 °C for 2 min, 95 °C for 2 min, and 40 cycles of 95 °C for 15 s followed by 60 °C for 1 min. A subsequent dissociation curve measurement from 60 °C to 95 °C was carried out. All samples were run in triplicate. PCR data were analyzed using the 7900 SDS v2.4.1 software (Applied Biosystems, Waltham, MA, USA). Relative gene expression was quantified by performing double delta Ct analysis ($2^{-\Delta\Delta Ct}$). B-actin Ct values were used as internal controls, and the gene of interest (GOI) expression was normalized to CD4+ cell expression.

### *4.4. Preparation of Fluorescently Labeled pETX*

pETX was provided by BEI (at a minimum >95% purity at 0.5 mg/mL (Epsilon Protoxin, from Clostridium perfringens, Strain 34 Type B, NR-856). pETX was labeled with Alexa Fluor 647 Protein Labeling Kit (Life Technologies, Carlsbad, CA, USA) per manufacturer's instructions. Labeled toxin was stored in a 50% glycerol stock (10 uM) at −20 °C until use.

### *4.5. Activation of ETX*

pETX provided by BEI was activated in house using immobilized trypsin, TPCK Treated, agarose resin (Thermo Fischer Scientific, Waltham, MA, USA). Briefly, 125 µL resin was washed three times in sodium phosphate buffer (pH 7.98). Resin was suspended in 200 µL sodium phosphate buffer and combined with 500 uL of BEI pETX (0.5 mg/mL) for two hours at 37 °C with gentle agitation. The solution was centrifuged at 18,000 rcf for 10 min, and the supernatant containing the activated ETX was collected. ETX activation was confirmed by the treatment of rMAL-CHO cells [7] to in-house controls. Activated toxin (~11 µM) was aliquoted and stored at −80 °C until use.

### *4.6. PBMNC Isolation from Human Peripheral Blood for ETX Binding and Cytotoxicity Studies*

Blood samples were collected from healthy controls via the cubital vein using BD Vacutainer K2 EDTA 7.2 mg Blood Collection tubes. Samples reached room temperature and were then diluted with an equal volume of Phosphate Buffered Saline (PBS) +2% Fetal Bovine Serum (FBS). Diluted blood was layered on top of Ficoll-Paque PLUS (GE Healthcare Bio-Sciences, Uppsala, Sweden). Tubes were centrifuged at 1200 rcf for 20 min at room temperature (without brakes). Buffy coat containing peripheral blood mononuclear cells (PBMNCs) was collected using a sterile transfer pipet. Buffy coat was washed with 20 mL of PBS + 2% FBS and centrifuged at 250 rcf for 10 min at 4 °C. The supernatant containing platelets was removed by aspiration. Cell pellet was washed with 20 mL of PBS + 2% FBS and centrifuged at 500 rcf for 10 min at 4 °C. The supernatant was aspirated, and cells were re-suspended in CTS™ OpTmizer™ T Cell Expansion Media (Thermo Fischer Scientific, Waltham, MA, USA) supplemented with Glutamax and 5% FBS. Cells were enumerated using a hemocytometer and adjusted to $1.5 \times 10^6$ cells/mL. Cells were kept on ice until use.

### *4.7. Evaluation of pETX-647 Binding to Lymphocyte Subsets*

To determine ETX binding, PBMNC ($1.5 \times 10^6$ cells/mL) were incubated with pETX-647 at 0 nM, 1 nM, 5 nM, 10 nM, 25 nM, and 50 nM for 15, 30, 60, and 120 min at 37 °C. In select experiments, media containing 50 nM pETX-647 was pretreated with or without an anti-ETX antibody (JL004) [78] for 30 min before treating cells for 2 h. At selected time points, 100 uL of cells were transferred to round bottom plates containing PBS + 2% FBS and immediately washed with PBS to remove unbound pETX-647. Cells were centrifuged at 500 rcf for 5 min. Cells were resuspended in Cell Staining buffer (Biolegend, San Diego, CA, USA) containing 5% Human TruStain FcX™ Fc Receptor Blocking Solution (BD Bioscience) for 10 min. Cells were then probed with FITC conjugated anti-CD4 Multiclone SK3 and SK4 (Biolegend, San Diego, CA, USA), PE-conjugated anti-CD8β Clone 2ST8.5H7 (BD Bioscience, San Diego, CA, USA), and V450 conjugated anti-CD19 clone SJ25C1 (BD Bioscience, San Jose, CA, USA) for 20 min at room temperature. Cells were washed and resuspended in PBS and analyzed using a BD FACSVerse Flow Cytometer (BD Bioscience, San Jose, CA, USA). From pETX-647 treatment to analysis via flow cytometry, cells were washed a total of three times. Data were collected using FACSuite™software (BD Bioscience, San Jose, CA, USA) and analyzed using FlowJo software (BD Bioscience, San Jose, CA, USA).

### *4.8. Evaluation of ETX-Induced Cytotoxicity in Lymphocyte Subsets*

To determine ETX-induced cytotoxicity, PBMNC ($1.5 \times 10^6$ cells/mL) were treated with activated ETX at 0 nM, 1 nM, 5 nM, 10 nM, 25 nM, and 50 nM for 30, 60, 120 and

240 min at 37 °C. In select experiments, media containing 50 nM ETX was pretreated with or without an anti-ETX antibody (JL008) [78] for 30 min before treating cells for 2 h. At selected time points, 100 uL of cells were transferred to round bottom plates containing ice cold PBS + 2%FBS to stop ETX activity [15]. Cells were immediately washed to remove unbound ETX and centrifuged at 500 rcf for 5 min at 4 °C. Cells were resuspended in Cell Staining buffer (Biolegend, San Diego, CA, USA) containing 5% Human TruStain FcX™ Fc Receptor Blocking Solution (BD Bioscience, San Jose, CA, USA) for 10 min. Cells were then probed with FITC conjugated anti-CD4 Multiclone SK3 and SK4 (Biolegend, San Diego, CA, USA), APC anti-CD8 clone SKI (Biolegend, San Diego, CA, USA), and V450 conjugated anti-CD19 clone SJ25C1 (BD Bioscience, San Jose, CA, USA) for 20 min at room temperature. Cells were washed and resuspended in PBS containing 2 μg/mL of PI and analyzed using a BD FACSVerse Flow Cytometer (BD Bioscience, San Jose, CA, USA). Data were collected using FACSuite™ software (BD Bioscience, San Jose, CA, USA) and analyzed using FlowJo software (BD Bioscience, San Jose, CA, USA).

### *4.9. Evalution of Pore Formation by Western Blot Analysis*

PBMNC ($1.5 \times 10^6$ cells/mL) were incubated with activated ETX at 0 nM, 10 nM, 25 nM, and 50 nM for 120 min at 37 °C. Alternatively, cells were treated with 50 nM of ETX for 30, 60, or 120 min. As a positive control for pore formation, rMAL-CHO cells treated with or without 50 nM active ETX for 30 min were used as controls [7,78]. After treatment, cells were immediately moved to ice, then washed three times with ice cold PBS. Cells were lysed in ice-cold RIPA buffer (50 mM Tris-HCl (pH 8.0), 150 mM NaCl, +1% NP-40, 0.1% Sodium dodecyl sulfate, 0.5% Sodium Deoxycholate) with proteinase and phosphatase inhibitors (Cell Signaling Technologies, Danvers, MA, USA) for 10 min. Samples were centrifuged at 5000 rcf for 5 min to pellet nuclei and DNA. Supernatants were collected and used for Western blot analysis. All samples were prepared in 2X Laemmli Sample Buffer (Bio-Rad, Hercules, CA, USA) containing 5% 2-Mercaptoethanol (Bio-Rad, Hercules, CA, USA) and heated at 95 °C for 5 min before loading onto 4–20% Mini-PROTEAN TGX Stain-Free gels (Bio-Rad, Hercules, CA, USA). Gels were run in Tris/Glycine SDS Buffer (Bio-Rad, Hercules, CA, USA) at 200 V for 35 min. Semi-dry transfers were performed in transfer Tris/Glycine Buffer (Bio-Rad, Hercules, CA, USA), using the Trans-Blot SD Semi-Dry Electrophoretic Transfer Cell system (Bio-Rad, Hercules, CA, USA) at 15 V for 15 min. Blots were blocked in 5% Blotting-Grade Blocker nonfat milk (Bio-Rad) in Tris Buffered Saline with Tween 20 (TBS-T, Cell Signaling Technology, Danvers, MA, USA) for one hour at room temperature. Blots were then incubated with primary antibodies anti-ETX antibody JL004 at 0.34 μg/mL in blocking solution overnight at 4 °C. Blots were washed three times for 5 min in TBS-T at room temperature and incubated with secondary antibody peroxidase-conjugated Affinipure Goat Anti-Rabbit IgG H + L (Jackson ImmunoResearch, Baltimore, MD, USA) at 0.024 μg/mL in blocking solution for 2 h at room temperature. Blots were washed three times for 5 min in TBS-T and developed for 5 min at room temperature in SuperSignal West Dura Extended Duration Substrate (ThermoFisher Scientific, Waltham, MA, USA). The developed blots were visualized on 5 × 7 CL-XPosure Films (ThermoFisher Scientific, Waltham, MA, USA) at various exposure times using a Konica Minolta SRX-101A film processor.

### *4.10. Statistics*

One-way ANOVA with post hoc Tukey HSD test was used to determine significance when comparing three or more data points. Unpaired Student's t-tests were used to determine significance when comparing only two data points. These instances are indicated in figure legends.

**Supplementary Materials:** The following are available online at https://www.mdpi.com/article/10.3390/toxins15070423/s1: Supplemental Table S1: *p* values of comparisons for different pETX doses for dose response curves [104]; Supplementary Figure S1: Confirmation of *Mal* Expression in primary human lymphocytes from publicly available datasets; Supplemental Figure S2: Gating strategy for lymphocyte populations; Supplemental Figure S3: STX binding to CD4+, CD8+, and CD19+ cells; Supplemental Figure S4: Additional time points for dose response evaluation in lymphocyte subsets.

**Author Contributions:** J.R.L. and T.V. conceived and designed experiments. P.W. and S.V.H. performed RT-qPCR experiments. J.R.L. and P.W. analyzed RT-qPCR data. J.R.L. and M.R.M. analyzed publicly available gene expression data. J.R.L. and S.V.S. performed flow cytometry experiments and analyzed data. J.R.L. and S.V.S. performed Western blots. K.R.R. and T.V. conceived the hypothesis for the paper. V.A.F. reviewed data and gave expert advice in experimental design and interpretation. J.R.L. wrote the paper and all authors assisted in editing. All authors have read and agreed to the published version of the manuscript.

**Funding:** This work was generously supported, in part, by NMSS awards RG-1508-06131.

**Institutional Review Board Statement:** Peripheral blood samples were collected from healthy controls in accordance with our Institutional Review Board, protocol number 1003010940.

**Informed Consent Statement:** Informed consent was obtained from all subjects involved in the study.

**Data Availability Statement:** The data presented in this study are available in this article and supplementary materials.

**Acknowledgments:** We are grateful to Yinghua Ma for his guidance with the RT-qPCR analysis. We are grateful to Kiel Telesford for his guidance with the flow cytometry analysis. All authors have read and agreed to the published version of the manuscript.

**Conflicts of Interest:** T.V., K.R.R., and V.A.F. are the inventors of US patent No. 9758573.

## References


1. Rumah, K.R.; Linden, J.; Fischetti, V.A.; Vartanian, T. Isolation of Clostridium perfringens type B in an individual at first clinical presentation of multiple sclerosis provides clues for environmental triggers of the disease. *PLoS ONE* **2013**, *8*, e76359. [CrossRef]
2. Wagley, S.; Bokori-Brown, M.; Morcrette, H.; Malaspina, A.; D'Arcy, C.; Gnanapavan, S.; Lewis, N.; Popoff, M.R.; Raciborska, D.; Nicholas, R.; et al. Evidence of Clostridium perfringens epsilon toxin associated with multiple sclerosis. *Mult. Scler.* **2018**, *25*, 653–660. [CrossRef] [PubMed]
3. Ma, Y.; Sannino, D.; Linden, J.R.; Haigh, S.; Zhao, B.; Grigg, J.B.; Zumbo, P.; Dundar, F.; Butler, D.; Profaci, C.P.; et al. Epsilon toxin-producing Clostridium perfringens colonize the multiple sclerosis gut microbiome overcoming CNS immune privilege. *J. Clin. Investig.* **2023**, *133*, e163239. [CrossRef] [PubMed]
4. Adler, D.; Linden, J.R.; Shetty, S.V.; Ma, Y.; Bokori-Brown, M.; Titball, R.W.; Vartanian, T. Clostridium perfringens Epsilon Toxin Compromises the Blood-Brain Barrier in a Humanized Zebrafish Model. *iScience* **2019**, *15*, 39–54. [CrossRef] [PubMed]
5. Linden, J.R.; Flores, C.; Schmidt, E.F.; Uzal, F.A.; Michel, A.O.; Valenzuela, M.; Dobrow, S.; Vartanian, T. Clostridium perfringens epsilon toxin induces blood brain barrier permeability via caveolae-dependent transcytosis and requires expression of MAL. *PLoS Pathog.* **2019**, *15*, e1008014. [CrossRef]
6. Linden, J.R.; Ma, Y.; Zhao, B.; Harris, J.M.; Rumah, K.R.; Schaeren-Wiemers, N.; Vartanian, T. Clostridium perfringens Epsilon Toxin Causes Selective Death of Mature Oligodendrocytes and Central Nervous System Demyelination. *mBio* **2015**, *6*, e02513. [CrossRef]
7. Rumah, K.R.; Ma, Y.; Linden, J.R.; Oo, M.L.; Anrather, J.; Schaeren-Wiemers, N.; Alonso, M.A.; Fischetti, V.A.; McClain, M.S.; Vartanian, T. The Myelin and Lymphocyte Protein MAL Is Required for Binding and Activity of Clostridium perfringens epsilon-Toxin. *PLoS Pathog.* **2015**, *11*, e1004896. [CrossRef]
8. Alonso, M.A.; Weissman, S.M. cDNA cloning and sequence of MAL, a hydrophobic protein associated with human T-cell differentiation. *Proc. Natl. Acad. Sci. USA* **1987**, *84*, 1997–2001. [CrossRef]
9. Blanch, M.; Dorca-Arevalo, J.; Not, A.; Cases, M.; Gomez de Aranda, I.; Martinez Yelamos, A.; Martinez Yelamos, S.; Solsona, C.; Blasi, J. The cytotoxicity of Epsilon toxin from Clostridium perfringens on lymphocytes is mediated by MAL protein expression. *Mol. Cell. Biol.* **2018**, *38*, e00086-18. [CrossRef]
10. Wang, R.; Zhang, P.; Wang, J.; Ma, L.; E, W.; Suo, S.; Jiang, M.; Li, J.; Chen, H.; Sun, H.; et al. Construction of a cross-species cell landscape at single-cell level. *Nucleic Acids Res.* **2023**, *51*, 501–516. [CrossRef]
11. van Langelaar, J.; Rijvers, L.; Smolders, J.; van Luijn, M.M. B and T Cells Driving Multiple Sclerosis: Identity, Mechanisms and Potential Triggers. *Front. Immunol.* **2020**, *11*, 760. [CrossRef]
12. Chitnis, T. The role of CD4 T cells in the pathogenesis of multiple sclerosis. *Int. Rev. Neurobiol.* **2007**, *79*, 43–72. [PubMed]

13. Bokori-Brown, M.; Savva, C.G.; Fernandes da Costa, S.P.; Naylor, C.E.; Basak, A.K.; Titball, R.W. Molecular basis of toxicity of Clostridium perfringens epsilon toxin. *FEBS J.* **2011**, *278*, 4589–4601. [CrossRef] [PubMed]
14. Popoff, M.R. Epsilon toxin: A fascinating pore-forming toxin. *FEBS J.* **2011**, *278*, 4602–4615. [CrossRef] [PubMed]
15. Robertson, S.L.; Li, J.; Uzal, F.A.; McClane, B.A. Evidence for a prepore stage in the action of Clostridium perfringens epsilon toxin. *PLoS ONE* **2011**, *6*, e22053. [CrossRef]
16. Savva, C.G.; Clark, A.R.; Naylor, C.E.; Popoff, M.R.; Moss, D.S.; Basak, A.K.; Titball, R.W.; Bokori-Brown, M. The pore structure of Clostridium perfringens epsilon toxin. *Nat. Commun.* **2019**, *10*, 2641. [CrossRef]
17. Garcia, J.P.; Adams, V.; Beingesser, J.; Hughes, M.L.; Poon, R.; Lyras, D.; Hill, A.; McClane, B.A.; Rood, J.I.; Uzal, F.A. Epsilon toxin is essential for the virulence of Clostridium perfringens type D infection in sheep, goats, and mice. *Infect. Immun.* **2013**, *81*, 2405–2414. [CrossRef]
18. Uzal, F.A.; Pasini, M.I.; Olaechea, F.V.; Robles, C.A.; Elizondo, A. An outbreak of enterotoxaemia caused by Clostridium perfringens type D in goats in Patagonia. *Vet. Rec.* **1994**, *135*, 279–280. [CrossRef]
19. Nagahama, M.; Sakurai, J. Distribution of labeled Clostridium perfringens epsilon toxin in mice. *Toxicon* **1991**, *29*, 211–217. [CrossRef]
20. Tamai, E.; Ishida, T.; Miyata, S.; Matsushita, O.; Suda, H.; Kobayashi, S.; Sonobe, H.; Okabe, A. Accumulation of Clostridium perfringens epsilon-toxin in the mouse kidney and its possible biological significance. *Infect. Immun.* **2003**, *71*, 5371–5375. [CrossRef]
21. Finnie, J.W. Pathogenesis of brain damage produced in sheep by Clostridium perfringens type D epsilon toxin: A review. *Aust. Vet. J.* **2003**, *81*, 219–221. [CrossRef]
22. Soler-Jover, A.; Blasi, J.; Gomez de Aranda, I.; Navarro, P.; Gibert, M.; Popoff, M.R.; Martin-Satue, M. Effect of epsilon toxin-GFP on MDCK cells and renal tubules in vivo. *J. Histochem. Cytochem.* **2004**, *52*, 931–942. [CrossRef] [PubMed]
23. Fernandez Miyakawa, M.E.; Zabal, O.; Silberstein, C. Clostridium perfringens epsilon toxin is cytotoxic for human renal tubular epithelial cells. *Hum. Exp. Toxicol.* **2011**, *30*, 275–282. [CrossRef]
24. Dorca-Arevalo, J.; Martin-Satue, M.; Blasi, J. Characterization of the high affinity binding of epsilon toxin from Clostridium perfringens to the renal system. *Vet. Microbiol.* **2012**, *157*, 179–189. [CrossRef]
25. Finnie, J.W. Neurological disorders produced by Clostridium perfringens type D epsilon toxin. *Anaerobe* **2004**, *10*, 145–150. [CrossRef] [PubMed]
26. Dorca-Arevalo, J.; Soler-Jover, A.; Gibert, M.; Popoff, M.R.; Martin-Satue, M.; Blasi, J. Binding of epsilon-toxin from Clostridium perfringens in the nervous system. *Vet. Microbiol.* **2008**, *131*, 14–25. [CrossRef] [PubMed]
27. Wioland, L.; Dupont, J.L.; Doussau, F.; Gaillard, S.; Heid, F.; Isope, P.; Pauillac, S.; Popoff, M.R.; Bossu, J.L.; Poulain, B. Epsilon toxin from Clostridium perfringens acts on oligodendrocytes without forming pores, and causes demyelination. *Cell. Microbiol.* **2015**, *17*, 369–388. [CrossRef]
28. Lonchamp, E.; Dupont, J.L.; Wioland, L.; Courjaret, R.; Mbebi-Liegeois, C.; Jover, E.; Doussau, F.; Popoff, M.R.; Bossu, J.L.; de Barry, J.; et al. Clostridium perfringens epsilon toxin targets granule cells in the mouse cerebellum and stimulates glutamate release. *PLoS ONE* **2010**, *5*, e13046. [CrossRef]
29. Worthington, R.W.; Mulders, M.S. Effect of Clostridium perfringens epsilon toxin on the blood brain barrier of mice. *Onderstepoort J. Vet. Res.* **1975**, *42*, 25–27.
30. Soler-Jover, A.; Dorca, J.; Popoff, M.R.; Gibert, M.; Saura, J.; Tusell, J.M.; Serratosa, J.; Blasi, J.; Martin-Satue, M. Distribution of Clostridium perfringens epsilon toxin in the brains of acutely intoxicated mice and its effect upon glial cells. *Toxicon* **2007**, *50*, 530–540. [CrossRef]
31. Fennessey, C.M.; Sheng, J.; Rubin, D.H.; McClain, M.S. Oligomerization of Clostridium perfringens epsilon toxin is dependent upon caveolins 1 and 2. *PLoS ONE* **2012**, *7*, e46866. [CrossRef] [PubMed]
32. Ivie, S.E.; Fennessey, C.M.; Sheng, J.; Rubin, D.H.; McClain, M.S. Gene-trap mutagenesis identifies mammalian genes contributing to intoxication by Clostridium perfringens epsilon-toxin. *PLoS ONE* **2011**, *6*, e17787. [CrossRef] [PubMed]
33. Anton, O.; Batista, A.; Millan, J.; Andres-Delgado, L.; Puertollano, R.; Correas, I.; Alonso, M.A. An essential role for the MAL protein in targeting Lck to the plasma membrane of human T lymphocytes. *J. Exp. Med.* **2008**, *205*, 3201–3213. [CrossRef]
34. Anton, O.M.; Andres-Delgado, L.; Reglero-Real, N.; Batista, A.; Alonso, M.A. MAL protein controls protein sorting at the supramolecular activation cluster of human T lymphocytes. *J. Immunol.* **2011**, *186*, 6345–6356. [CrossRef]
35. Millan, J.; Alonso, M.A. MAL, a novel integral membrane protein of human T lymphocytes, associates with glycosylphosphatidylinositol-anchored proteins and Src-like tyrosine kinases. *Eur. J. Immunol.* **1998**, *28*, 3675–3684. [CrossRef]
36. Millan, J.; Puertollano, R.; Fan, L.; Alonso, M.A. Caveolin and MAL, two protein components of internal detergent-insoluble membranes, are in distinct lipid microenvironments in MDCK cells. *Biochem. Biophys. Res. Commun.* **1997**, *233*, 707–712. [CrossRef]
37. Millan, J.; Puertollano, R.; Fan, L.; Rancano, C.; Alonso, M.A. The MAL proteolipid is a component of the detergent-insoluble membrane subdomains of human T-lymphocytes. *Biochem. J.* **1997**, *321 Pt 1*, 247–252. [CrossRef] [PubMed]
38. Ventimiglia, L.N.; Fernandez-Martin, L.; Martinez-Alonso, E.; Anton, O.M.; Guerra, M.; Martinez-Menarguez, J.A.; Andres, G.; Alonso, M.A. Cutting Edge: Regulation of Exosome Secretion by the Integral MAL Protein in T Cells. *J. Immunol.* **2015**, *195*, 810–814. [CrossRef]

39. Frank, M. MAL, a proteolipid in glycosphingolipid enriched domains: Functional implications in myelin and beyond. *Prog. Neurobiol.* **2000**, *60*, 531–544. [CrossRef]
40. Schaeren-Wiemers, N.; Schaefer, C.; Valenzuela, D.M.; Yancopoulos, G.D.; Schwab, M.E. Identification of new oligodendrocyte- and myelin-specific genes by a differential screening approach. *J. Neurochem.* **1995**, *65*, 10–22. [CrossRef]
41. Schaeren-Wiemers, N.; Valenzuela, D.M.; Frank, M.; Schwab, M.E. Characterization of a rat gene, rMAL, encoding a protein with four hydrophobic domains in central and peripheral myelin. *J. Neurosci.* **1995**, *15*, 5753–5764. [CrossRef] [PubMed]
42. Zacchetti, D.; Peranen, J.; Murata, M.; Fiedler, K.; Simons, K. VIP17/MAL, a proteolipid in apical transport vesicles. *FEBS Lett.* **1995**, *377*, 465–469. [PubMed]
43. Kim, T.; Fiedler, K.; Madison, D.L.; Krueger, W.H.; Pfeiffer, S.E. Cloning and characterization of MVP17: A developmentally regulated myelin protein in oligodendrocytes. *J. Neurosci. Res.* **1995**, *42*, 413–422. [CrossRef]
44. Caduff, J.; Sansano, S.; Bonnet, A.; Suter, U.; Schaeren-Wiemers, N. Characterization of GFP-MAL expression and incorporation in rafts. *Microsc. Res. Tech.* **2001**, *52*, 645–655. [CrossRef] [PubMed]
45. Magal, L.G.; Yaffe, Y.; Shepshelovich, J.; Aranda, J.F.; de Marco Mdel, C.; Gaus, K.; Alonso, M.A.; Hirschberg, K. Clustering and lateral concentration of raft lipids by the MAL protein. *Mol. Biol. Cell* **2009**, *20*, 3751–3762. [CrossRef]
46. Ramnarayanan, S.P.; Tuma, P.L. MAL, but not MAL2, expression promotes the formation of cholesterol-dependent membrane domains that recruit apical proteins. *Biochem. J.* **2011**, *439*, 497–504. [CrossRef]
47. Alonso, M.A.; Millan, J. The role of lipid rafts in signalling and membrane trafficking in T lymphocytes. *J. Cell. Sci.* **2001**, *114 Pt 22*, 3957–3965. [CrossRef]
48. Martin-Belmonte, F.; Arvan, P.; Alonso, M.A. MAL mediates apical transport of secretory proteins in polarized epithelial Madin-Darby canine kidney cells. *J. Biol. Chem.* **2001**, *276*, 49337–49342. [CrossRef]
49. Martin-Belmonte, F.; Puertollano, R.; Millan, J.; Alonso, M.A. The MAL proteolipid is necessary for the overall apical delivery of membrane proteins in the polarized epithelial Madin-Darby canine kidney and fischer rat thyroid cell lines. *Mol. Biol. Cell* **2000**, *11*, 2033–2045. [CrossRef]
50. Puertollano, R.; Alonso, M.A. MAL, an integral element of the apical sorting machinery, is an itinerant protein that cycles between the trans-Golgi network and the plasma membrane. *Mol. Biol. Cell* **1999**, *10*, 3435–3447. [CrossRef]
51. Puertollano, R.; Martin-Belmonte, F.; Millan, J.; de Marco, M.C.; Albar, J.P.; Kremer, L.; Alonso, M.A. The MAL proteolipid is necessary for normal apical transport and accurate sorting of the influenza virus hemagglutinin in Madin-Darby canine kidney cells. *J. Cell Biol.* **1999**, *145*, 141–151. [CrossRef] [PubMed]
52. Puertollano, R.; Martinez-Menarguez, J.A.; Batista, A.; Ballesta, J.; Alonso, M.A. An intact dilysine-like motif in the carboxyl terminus of MAL is required for normal apical transport of the influenza virus hemagglutinin cargo protein in epithelial Madin-Darby canine kidney cells. *Mol. Biol. Cell* **2001**, *12*, 1869–1883. [CrossRef] [PubMed]
53. Puertollano, R.; Menendez, M.; Alonso, M.A. Incorporation of MAL, an integral protein element of the machinery for the glycolipid and cholesterol-mediated apical pathway of transport, into artificial membranes requires neither of these lipid species. *Biochem. Biophys. Res. Commun.* **1999**, *266*, 330–333. [CrossRef] [PubMed]
54. Schaeren-Wiemers, N.; Bonnet, A.; Erb, M.; Erne, B.; Bartsch, U.; Kern, F.; Mantei, N.; Sherman, D.; Suter, U. The raft-associated protein MAL is required for maintenance of proper axon--glia interactions in the central nervous system. *J. Cell Biol.* **2004**, *166*, 731–742. [CrossRef]
55. Andres-Delgado, L.; Anton, O.M.; Madrid, R.; Byrne, J.A.; Alonso, M.A. Formin INF2 regulates MAL-mediated transport of Lck to the plasma membrane of human T lymphocytes. *Blood* **2010**, *116*, 5919–5929. [CrossRef]
56. Copie-Bergman, C.; Plonquet, A.; Alonso, M.A.; Boulland, M.L.; Marquet, J.; Divine, M.; Moller, P.; Leroy, K.; Gaulard, P. MAL expression in lymphoid cells: Further evidence for MAL as a distinct molecular marker of primary mediastinal large B-cell lymphomas. *Mod. Pathol.* **2002**, *15*, 1172–1180. [CrossRef]
57. Rancano, C.; Rubio, T.; Alonso, M.A. Alternative splicing of human T-cell-specific MAL mRNA and its correlation with the exon/intron organization of the gene. *Genomics* **1994**, *21*, 447–450. [CrossRef]
58. Rancano, C.; Rubio, T.; Correas, I.; Alonso, M.A. Genomic structure and subcellular localization of MAL, a human T-cell-specific proteolipid protein. *J. Biol. Chem.* **1994**, *269*, 8159–8164. [CrossRef]
59. Gentry, M.; Bodo, J.; Durkin, L.; Hsi, E.D. Performance of a Commercially Available MAL Antibody in the Diagnosis of Primary Mediastinal Large B-Cell Lymphoma. *Am. J. Surg. Pathol.* **2017**, *41*, 189–194. [CrossRef]
60. Kohno, T.; Moriuchi, R.; Katamine, S.; Yamada, Y.; Tomonaga, M.; Matsuyama, T. Identification of genes associated with the progression of adult T cell leukemia (ATL). *Jpn. J. Cancer Res.* **2000**, *91*, 1103–1110. [CrossRef]
61. Lara-Lemus, R. On The Role of Myelin and Lymphocyte Protein (MAL) In Cancer: A Puzzle With Two Faces. *J. Cancer* **2019**, *10*, 2312–2318. [CrossRef] [PubMed]
62. Rubio-Ramos, A.; Labat-de-Hoz, L.; Correas, I.; Alonso, M.A. The MAL Protein, an Integral Component of Specialized Membranes, in Normal Cells and Cancer. *Cells* **2021**, *10*, 1065. [CrossRef] [PubMed]
63. Chan, J.K. Mediastinal large B-cell lymphoma: New evidence in support of its distinctive identity. *Adv. Anat. Pathol.* **2000**, *7*, 201–209. [CrossRef]
64. Copie-Bergman, C.; Gaulard, P.; Maouche-Chretien, L.; Briere, J.; Haioun, C.; Alonso, M.A.; Romeo, P.H.; Leroy, K. The MAL gene is expressed in primary mediastinal large B-cell lymphoma. *Blood* **1999**, *94*, 3567–3575. [CrossRef] [PubMed]

65. Hsi, E.D.; Sup, S.J.; Alemany, C.; Tso, E.; Skacel, M.; Elson, P.; Alonso, M.A.; Pohlman, B. MAL is expressed in a subset of Hodgkin lymphoma and identifies a population of patients with poor prognosis. *Am. J. Clin. Pathol.* **2006**, *125*, 776–782. [CrossRef]
66. Zheng, G.X.; Terry, J.M.; Belgrader, P.; Ryvkin, P.; Bent, Z.W.; Wilson, R.; Ziraldo, S.B.; Wheeler, T.D.; McDermott, G.P.; Zhu, J.; et al. Massively parallel digital transcriptional profiling of single cells. *Nat. Commun.* **2017**, *8*, 14049. [CrossRef] [PubMed]
67. Uhlen, M.; Karlsson, M.J.; Zhong, W.; Tebani, A.; Pou, C.; Mikes, J.; Lakshmikanth, T.; Forsstrom, B.; Edfors, F.; Odeberg, J.; et al. A genome-wide transcriptomic analysis of protein-coding genes in human blood cells. *Science* **2019**, *366*, eaax9198. [CrossRef]
68. Watkins, N.A.; Gusnanto, A.; de Bono, B.; De, S.; Miranda-Saavedra, D.; Hardie, D.L.; Angenent, W.G.; Attwood, A.P.; Ellis, P.D.; Erber, W.; et al. A HaemAtlas: Characterizing gene expression in differentiated human blood cells. *Blood* **2009**, *113*, e1–e9. [CrossRef]
69. Choi, J.; Baldwin, T.M.; Wong, M.; Bolden, J.E.; Fairfax, K.A.; Lucas, E.C.; Cole, R.; Biben, C.; Morgan, C.; Ramsay, K.A.; et al. Haemopedia RNA-seq: A database of gene expression during haematopoiesis in mice and humans. *Nucleic Acids Res.* **2019**, *47*, D780–D785. [CrossRef]
70. Nagahama, M.; Ochi, S.; Sakurai, J. Assembly of Clostridium perfringens epsilon-toxin on MDCK cell membrane. *J. Nat. Toxins* **1998**, *7*, 291–302.
71. Petit, L.; Gibert, M.; Gillet, D.; Laurent-Winter, C.; Boquet, P.; Popoff, M.R. Clostridium perfringens epsilon-toxin acts on MDCK cells by forming a large membrane complex. *J. Bacteriol.* **1997**, *179*, 6480–6487. [CrossRef] [PubMed]
72. Dorca-Arevalo, J.; Gomez de Aranda, I.; Blasi, J. New Mutants of Epsilon Toxin from Clostridium perfringens with an Altered Receptor-Binding Site and Cell-Type Specificity. *Toxins* **2022**, *14*, 288. [CrossRef]
73. Miyata, S.; Matsushita, O.; Minami, J.; Katayama, S.; Shimamoto, S.; Okabe, A. Cleavage of a C-terminal peptide is essential for heptamerization of Clostridium perfringens epsilon-toxin in the synaptosomal membrane. *J. Biol. Chem.* **2001**, *276*, 13778–13783. [CrossRef]
74. Miyata, S.; Minami, J.; Tamai, E.; Matsushita, O.; Shimamoto, S.; Okabe, A. Clostridium perfringens epsilon-toxin forms a heptameric pore within the detergent-insoluble microdomains of Madin-Darby canine kidney cells and rat synaptosomes. *J. Biol. Chem.* **2002**, *277*, 39463–39468. [CrossRef]
75. Llorente, A.; de Marco, M.C.; Alonso, M.A. Caveolin-1 and MAL are located on prostasomes secreted by the prostate cancer PC-3 cell line. *J. Cell Sci.* **2004**, *117 Pt 22*, 5343–5351. [CrossRef] [PubMed]
76. Nagahama, M.; Itohayashi, Y.; Hara, H.; Higashihara, M.; Fukatani, Y.; Takagishi, T.; Oda, M.; Kobayashi, K.; Nakagawa, I.; Sakurai, J. Cellular vacuolation induced by Clostridium perfringens epsilon-toxin. *FEBS J.* **2011**, *278*, 3395–3407. [CrossRef] [PubMed]
77. Ventimiglia, L.N.; Alonso, M.A. Biogenesis and Function of T Cell-Derived Exosomes. *Front. Cell Dev. Biol.* **2016**, *4*, 84. [CrossRef]
78. Linden, J.R.; Telesford, K.; Shetty, S.; Winokour, P.; Haigh, S.; Cahir-McFarland, E.; Antognetti, G.; Datta, A.; Wang, T.; Meier, W.; et al. A Novel Panel of Rabbit Monoclonal Antibodies and Their Diverse Applications Including Inhibition of Clostridium perfringens Epsilon Toxin Oligomerization. *Antibodies* **2018**, *7*, 37. [CrossRef]
79. Kouzel, I.U.; Pohlentz, G.; Storck, W.; Radamm, L.; Hoffmann, P.; Bielaszewska, M.; Bauwens, A.; Cichon, C.; Schmidt, M.A.; Mormann, M.; et al. Association of Shiga toxin glycosphingolipid receptors with membrane microdomains of toxin-sensitive lymphoid and myeloid cells. *J. Lipid Res.* **2013**, *54*, 692–710. [CrossRef]
80. Khan, M.I.; Czarnecka, A.M.; Lewicki, S.; Helbrecht, I.; Brodaczewska, K.; Koch, I.; Zdanowski, R.; Krol, M.; Szczylik, C. Comparative Gene Expression Profiling of Primary and Metastatic Renal Cell Carcinoma Stem Cell-Like Cancer Cells. *PLoS ONE* **2016**, *11*, e0165718. [CrossRef]
81. Martin-Belmonte, F.; Kremer, L.; Albar, J.P.; Marazuela, M.; Alonso, M.A. Expression of the MAL gene in the thyroid: The MAL proteolipid, a component of glycolipid-enriched membranes, is apically distributed in thyroid follicles. *Endocrinology* **1998**, *139*, 2077–2084. [CrossRef]
82. Palacios, E.H.; Weiss, A. Function of the Src-family kinases, Lck and Fyn, in T-cell development and activation. *Oncogene* **2004**, *23*, 7990–8000. [CrossRef]
83. Pierce, S.K. Lipid rafts and B-cell activation. *Nat. Rev. Immunol.* **2002**, *2*, 96–105. [CrossRef]
84. Dorfman, D.M.; Shahsafaei, A.; Alonso, M.A. Utility of CD200 immunostaining in the diagnosis of primary mediastinal large B cell lymphoma: Comparison with MAL, CD23, and other markers. *Mod. Pathol.* **2012**, *25*, 1637–1643. [CrossRef] [PubMed]
85. Petit, L.; Gibert, M.; Gourch, A.; Bens, M.; Vandewalle, A.; Popoff, M.R. Clostridium perfringens epsilon toxin rapidly decreases membrane barrier permeability of polarized MDCK cells. *Cell. Microbiol.* **2003**, *5*, 155–164. [CrossRef] [PubMed]
86. Chassin, C.; Bens, M.; de Barry, J.; Courjaret, R.; Bossu, J.L.; Cluzeaud, F.; Ben Mkaddem, S.; Gibert, M.; Poulain, B.; Popoff, M.R.; et al. Pore-forming epsilon toxin causes membrane permeabilization and rapid ATP depletion-mediated cell death in renal collecting duct cells. *Am. J. Physiol. Renal Physiol.* **2007**, *293*, F927–F937. [CrossRef] [PubMed]
87. Kang, J.; Gao, J.; Yao, W.; Kang, L.; Gao, S.; Yang, H.; Ji, B.; Li, P.; Liu, J.; Yao, J.; et al. F199E substitution reduced toxicity of Clostridium perfringens epsilon toxin by depriving the receptor binding capability. *Hum. Vaccines Immunother.* **2017**, *13*, 1598–1608. [CrossRef]
88. Frischer, J.M.; Weigand, S.D.; Guo, Y.; Kale, N.; Parisi, J.E.; Pirko, I.; Mandrekar, J.; Bramow, S.; Metz, I.; Bruck, W.; et al. Clinical and pathological insights into the dynamic nature of the white matter multiple sclerosis plaque. *Ann. Neurol.* **2015**, *78*, 710–721. [CrossRef]

89. Kuhlmann, T.; Ludwin, S.; Prat, A.; Antel, J.; Bruck, W.; Lassmann, H. An updated histological classification system for multiple sclerosis lesions. *Acta Neuropathol.* **2017**, *133*, 13–24. [CrossRef]
90. Ramaglia, V.; Sheikh-Mohamed, S.; Legg, K.; Park, C.; Rojas, O.L.; Zandee, S.; Fu, F.; Ornatsky, O.; Swanson, E.C.; Pitt, D.; et al. Multiplexed imaging of immune cells in staged multiple sclerosis lesions by mass cytometry. *Elife* **2019**, *8*, e48051. [CrossRef]
91. Sospedra, M.; Martin, R. Immunology of Multiple Sclerosis. *Semin. Neurol.* **2016**, *36*, 115–127. [CrossRef] [PubMed]
92. Sospedra, M.; Martin, R. Immunology of multiple sclerosis. *Annu. Rev. Immunol.* **2005**, *23*, 683–747. [CrossRef] [PubMed]
93. Lucchinetti, C.; Bruck, W.; Parisi, J.; Scheithauer, B.; Rodriguez, M.; Lassmann, H. Heterogeneity of multiple sclerosis lesions: Implications for the pathogenesis of demyelination. *Ann. Neurol.* **2000**, *47*, 707–717. [CrossRef] [PubMed]
94. Bischof, L.J.; Kao, C.Y.; Los, F.C.; Gonzalez, M.R.; Shen, Z.; Briggs, S.P.; van der Goot, F.G.; Aroian, R.V. Activation of the unfolded protein response is required for defenses against bacterial pore-forming toxin in vivo. *PLoS Pathog.* **2008**, *4*, e1000176. [CrossRef] [PubMed]
95. Bischofberger, M.; Iacovache, I.; van der Goot, F.G. Pathogenic pore-forming proteins: Function and host response. *Cell Host Microbe* **2012**, *12*, 266–275. [CrossRef]
96. Chen, C.S.; Bellier, A.; Kao, C.Y.; Yang, Y.L.; Chen, H.D.; Los, F.C.; Aroian, R.V. WWP-1 is a novel modulator of the DAF-2 insulin-like signaling network involved in pore-forming toxin cellular defenses in Caenorhabditis elegans. *PLoS ONE* **2010**, *5*, e9494. [CrossRef]
97. Gonzalez, M.R.; Bischofberger, M.; Freche, B.; Ho, S.; Parton, R.G.; van der Goot, F.G. Pore-forming toxins induce multiple cellular responses promoting survival. *Cell. Microbiol.* **2011**, *13*, 1026–1043. [CrossRef]
98. Gurcel, L.; Abrami, L.; Girardin, S.; Tschopp, J.; van der Goot, F.G. Caspase-1 activation of lipid metabolic pathways in response to bacterial pore-forming toxins promotes cell survival. *Cell* **2006**, *126*, 1135–1145. [CrossRef]
99. Huffman, D.L.; Abrami, L.; Sasik, R.; Corbeil, J.; van der Goot, F.G.; Aroian, R.V. Mitogen-activated protein kinase pathways defend against bacterial pore-forming toxins. *Proc. Natl. Acad. Sci. USA* **2004**, *101*, 10995–11000. [CrossRef]
100. Kao, C.Y.; Los, F.C.; Huffman, D.L.; Wachi, S.; Kloft, N.; Husmann, M.; Karabrahimi, V.; Schwartz, J.L.; Bellier, A.; Ha, C.; et al. Global functional analyses of cellular responses to pore-forming toxins. *PLoS Pathog.* **2011**, *7*, e1001314. [CrossRef]
101. Kennedy, C.L.; Smith, D.J.; Lyras, D.; Chakravorty, A.; Rood, J.I. Programmed cellular necrosis mediated by the pore-forming alpha-toxin from Clostridium septicum. *PLoS Pathog.* **2009**, *5*, e1000516. [CrossRef]
102. Wiles, T.J.; Dhakal, B.K.; Eto, D.S.; Mulvey, M.A. Inactivation of host Akt/protein kinase B signaling by bacterial pore-forming toxins. *Mol. Biol. Cell* **2008**, *19*, 1427–1438. [CrossRef] [PubMed]
103. Verma, P.; Gandhi, S.; Lata, K.; Chattopadhyay, K. Pore-forming toxins in infection and immunity. *Biochem. Soc. Trans.* **2021**, *49*, 455–465. [CrossRef] [PubMed]
104. Stuart, T.; Butler, A.; Hoffman, P.; Hafemeister, C.; Papalexi, E.; Mauck, W.M., 3rd; Hao, Y.; Stoeckius, M.; Smibert, P.; Satija, R. Comprehensive Integration of Single-Cell Data. *Cell* **2019**, *177*, 1888–1902.e21. [CrossRef] [PubMed]

*toxins*

MDPI

*Article*

# PLGA Nanoparticle-Based Dissolving Microneedle Vaccine of *Clostridium perfringens* ε Toxin

**Wei Wan [1,2,†], Yue Li [1,†], Jing Wang [1], Zhiying Jin [1,3], Wenwen Xin [1], Lin Kang [1], Junhong Wang [4], Xiaoyang Li [1,2], Yakun Cao [1,3], Hao Yang [5,†], Jinglin Wang [1,*] and Shan Gao [1,*]**

1 State Key Laboratory of Pathogen and Biosecurity, Institute of Microbiology and Epidemiology, Academy of Military Medical Sciences (AMMS), Beijing 100071, China; wanwan03012022@163.com (W.W.); lyzaokewenhua@163.com (Y.L.); amms_wj@163.com (J.W.); jinzhiying96@163.com (Z.J.); xinww@hotmail.com (W.X.); kang_lin@hotmail.com (L.K.); lxy2563922462@163.com (X.L.); caoyakun0604@163.com (Y.C.)
2 College of Life Science and Technology, Beijing University of Chemical Technology, Beijing 100029, China
3 Hebei Key Laboratory of Animal Physiology, Biochemistry and Molecular Biology, College of Life Sciences, Hebei Normal University, Shijiazhuang 050024, China
4 School of Medical Technology, Tianjin University of Traditional Chinese Medicine, Tianjin 301617, China; wangjunhong11007@163.com
5 Beijing Noninvasion Biomedical Technology Co., Ltd., Beijing 101111, China; iam@noninvasion.com
* Correspondence: wangjlin@bmi.ac.cn (J.W.); gaoshan@bmi.ac.cn (S.G.); Tel./Fax: +86-010-66948531 (J.W.); +86-010-66948643 (S.G.)
† These authors contributed equally to this work.

**Abstract:** Epsilon toxin (ETX) is an exotoxin produced by type B and D *Clostridium perfringens* that causes enterotoxemia or necrotic enteritis in animals such as goats, sheep, and cattle. Vaccination is a key method in preventing such diseases. In this study, we developed a new type of dissolving microneedle patch (dMN) with a nanoparticle adjuvant for enhanced immune response to deliver the rETX$^{Y196E}$-C protein vaccine. We chose FDA-approved poly(lactic-co-glycolic acid) (PLGA) to prepare nanospheres as the vaccine adjuvant and introduced dimethyldioctadecylammonium bromide (DDAB) to make the surface of PLGA nanoparticles (PLGA NPs) positively charged for antigen adsorption. PLGA NPs with a diameter of 100~200 nm, a surface ZETA potential of approximately +40 mV, and good safety were successfully prepared and could effectively adsorb rETX$^{Y196E}$-C protein. Using non-toxic and antibacterial fish gelatin as the microneedle (MN) matrix, we prepared a PLGA-DDAB dMN vaccine with good mechanical properties that successfully penetrated the skin. After immunization of subcutaneous (SC) and dMN, antibody titers of the PLGA and Al adjuvant groups were similar in both two immune ways. However, in vivo neutralization experiments showed that the dMN vaccines had a better protective effect. When challenged with 100 × $LD_{50}$ GST-ETX, the survival rate of the MN group was 100%, while that of the SC Al group was 80%. However, a 100% protective effect was achieved in both immunization methods using PLGA NPs. In vitro neutralization experiments showed that the serum antibodies from the dMN and SC PLGA NPs groups both protect naive mice from 10 × $LD_{50}$ GST-ETX attack after being diluted 20 times and could also protect MDCK cells from 20 × $CT_{50}$ GST-ETX attack. In conclusion, the PLGA-DDAB dMN vaccine we prepared has good mechanical properties, immunogenicity, and protection, and can effectively prevent ETX poisoning. This provides a better way of delivering protein vaccines.

**Keywords:** *Clostridium perfringens* epsilon toxin; dissolving microneedle patch; poly(lactic-co-glycolic acid; vaccine adjuvants

**Key Contribution:** This study is the first to successfully develop a PLGA nanoparticle-based dissolving microneedle to deliver a vaccine against *Clostridium perfringens* ε toxin. It has good mechanical properties, immunogenicity and protection, and effectively prevents ETX poisoning. This provides an improved method for delivering protein vaccines.

## 1. Introduction

Epsilon toxin (ETX) is a pathogenic factor produced by *Clostridium perfringens* types B and D that can cause enterotoxemia or necrotizing enteritis in lambs and calves. Its toxicity is second only to that of botulinum and tetanus toxins, and it is classified as an important biological terrorist agent by the international community. ETX has the characteristics of rapid onset, short course, and high mortality rate. It is difficult to achieve a therapeutic effect by medication alone after poisoning, and acute cases often result in death before medication can take effect, causing huge losses to the livestock industry. The potential hazards of ETX to humans and animals, as well as the threat of bioterrorist attacks worldwide, have accelerated the development of vaccines to significantly reduce the incidence and mortality of ETX poisoning [1].

Most traditional vaccines used to prevent ETX poisoning are detoxified protein vaccines that can cause inflammation and damage to animals, reducing their safety. Compared with traditional attenuated vaccines, recombinant protein vaccines have significant advantages such as convenient production, fast speed, simple purification, high yield, small side effects, and higher safety, and have gradually replaced traditional detoxified vaccines. However, due to their low molecular weight, poor immunogenicity, and rapid vaccine release, they cannot stimulate the production of long-lasting and effective immune responses [2]; aluminum (Al) adjuvants are typically needed to achieve the expected immune effect. Al adjuvants are widely used and universal vaccine adjuvants that can promote both Th2 humoral immune responses and cell-mediated immune responses [3]. However, they have side effects such as neurotoxicity, local fever, and swelling after injection; thus, an urgent need for safe and effective new adjuvants exists [4,5]. Recently, biodegradable polymer particle adjuvants have received extensive attention, especially poly(lactic-co-glycolic acid) nanoparticles (PLGA NPs), which are FDA-approved pharmaceutical excipients [6] with high safety and biodegradability. PLGA NPs can adsorb or encapsulate antigens, acting as vaccine carriers, forming a vaccine storage depot, delivering antigens and prolonging their activity time. They can also enhance antigen uptake, activation, and maturation of dendritic cells (DC) [7], which can enhance cellular immunity as an immunological adjuvant. Researchers have used PLGA as an adjuvant to deliver protein antigens such as ovalbumin (OVA) [7,8], hepatitis B [9,10], malaria [11,12], and *Chlamydia trachomatis* [13,14], all of which have effectively and safely induced immune responses.

As a new vaccine delivery approach, microneedles (MNs) penetrate the stratum corneum with their micron-sized needle-like projections, delivering vaccines to Langerhans cells and dendritic cells in the epidermis and inducing an adaptive immune response [15,16]. Compared to traditional vaccination methods, MN vaccines have many advantages, including good heat stability, self-administration, almost painless administration, and no sharps waste [17–19]. By combining PLGA NPs containing antigens with MN delivery technology, a new type of immuno-enhanced MN is explored. Using PLGA NPs to protect antigens from environmental influences and enhance the permeability of nanoparticles in the skin [20], the long-term stability of antigens can be improved [21] and the sustained release of antigens can be extended [7] to enhance the immune effects of vaccine antigens. The ease and convenience of MN delivery make possible the self-administration of vaccines, thereby improving vaccination efficiency. By combining the advantages of the above two technologies, the preventive effect of vaccines can be enhanced, providing an effective means for preventing ETX toxicity. Zheng and co-workers have reported a successful case of combining PLGA NPs vaccines with MN to deliver hepatitis B protein vaccines to the skin [10].

In this study, we develop a dissolving microneedle patch (dMN) using a PLGA NPs adjuvant to deliver the rETX$^{Y196E}$-C protein vaccine. We introduce the addition of dimethyl-dioctadecylammonium bromide (DDAB) to make the surface of PLGA NPs positively charged for antigen adsorption, creating a PLGA-DDAB/rETX$^{Y196E}$-C dMN vaccine with significant advantages, including safety, high efficiency, and easy administration. The sustained-release effect of PLGA NPs as an adjuvant was evaluated and the immune response type and intensity stimulated by the dMN vaccine in mice were detected and analyzed. This effectively improved the problems of weak immunogenicity and inconvenient injection of traditional ETX recombinant protein vaccines.

## 2. Results

### *2.1. Preparation of Vaccine Antigens and Toxin Proteins*

We prepared attenuated mutant protein rETX$^{Y196E}$-C and GST-ETX recombinant protein using bacterial strains stored in our laboratory. The rETX$^{Y196E}$-C protein was obtained by substituting the 196th tyrosine (Y196) with glutamic acid (E) and introducing a C-terminal peptide consisting of 23 amino acids, which is a variant with low toxicity used for mouse immunization, while the GST-ETX protein retained toxicity and was used for immunization and subsequent challenge. We purified the proteins using nickel-nitrilotriacetic affinity chromatography and gradient elution. The purified rETX$^{Y196E}$-C protein had a molecular weight of 33 kDa (Figure 1A,B), while the purified GST-ETX protein had a molecular weight of 56 kDa (Figure 1C,D). The purity of both proteins was greater than 81% as determined by SDS-PAGE analysis. The concentrated protein was used for subsequent experiments. The concentration of rETX$^{Y196E}$-C and GST-ETX proteins were estimated as 31.08 mg/mL and 2.12 mg/mL, respectively, using a BCA protein quantification kit.

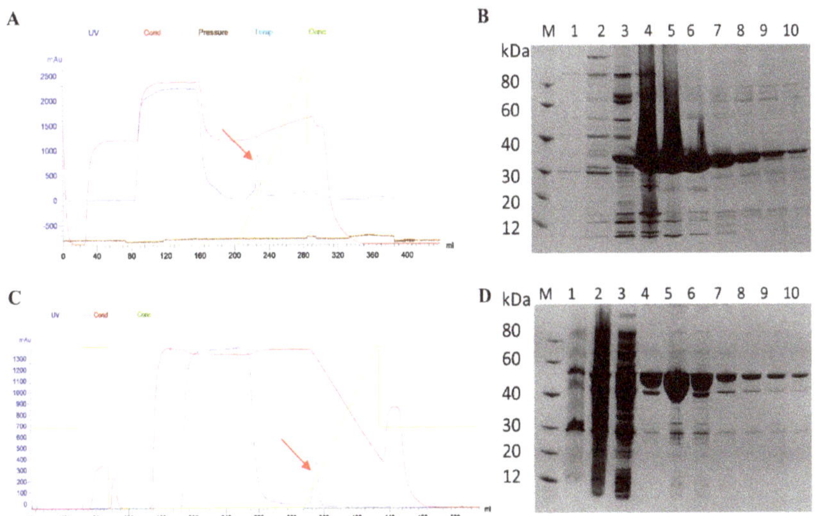


**Figure 1.** (**A**) Purification profile of rETX$^{Y196E}$-C protein, with the red arrow indicating the target protein peak. (**B**) SDS-PAGE analysis of rETX$^{Y196E}$-C protein. M: molecular weight marker; lanes 1–2: induced bacterial culture, sonicated pellet and supernatant; lane 3: flow-through; lanes 4–10: target protein. (**C**) Purification profile of GST−ETX protein, with the red arrow indicating the target protein peak. (**D**) SDS−PAGE analysis of GST-ETX protein. M: molecular weight marker; lanes 1–2: induced bacterial culture, sonicated pellet and supernatant; lane 3: flow-through; lanes 4–10: target protein.

### *2.2. Characterization of PLGA-DDAB Nanoparticles*

The surface of simple PLGA NPs is negatively charged and cannot directly adsorb negatively charged proteins. However, the introduction of a cationic lipid, DDAB, makes the surface of PLGA NPs positively charged, allowing the negatively charged antigen (Ag), rETX$^{Y196E}$-C protein, to be adsorbed onto their surface through electrostatic interactions (Figure 2A). PLGA-DDAB NPs were successfully prepared using the emulsion–evaporation method. SEM scanning electron microscopy (Figure 2B) and field emission transmission electron microscopy (Figure 2C) showed that the PLGA-DDAB NPs were spherical, smooth-surfaced, had a particle size of about 100 nm, and could be efficiently absorbed by antigen-presenting cells (APCs) to enhance immune responses [22]. The introduction of DDAB increased the particle size of

PLGA NPs by 12.71 ± 2.895 nm, indicating the successful introduction of DDAB into PLGA NPs (Figure 2E). To further verify this result, EDS spectral scanning analysis of the Br and N elements unique to DDAB was used to confirm the presence of both elements in PLGA NPs (Figure 2F,H). The amount of DDAB added significantly affected the ZETA potential of the PLGA surface (Figure 2D), which in turn affected the protein adsorption efficiency. Therefore, we added 0.005, 0.01, 0.02, or 0.03 g of DDAB to a 0.1 g PLGA solution and observed the effect of different amounts of DDAB on the ZETA potential of the particle surface. When 0.02 g of DDAB was added, the ZETA potential of PLGA-DDAB NPs reached the highest level. The ZETA potential and polymer dispersity index (PDI) can directly reflect the stability and dispersibility of the particles. The particle size distribution, ZETA potential, and PDI of the NPs were determined using a laser particle size analyzer. Each sample was measured three times and the mean value was used in the analyses (Table 1). The PDI of PLGA-DDAB was less than 0.1, indicating good stability and dispersibility of the particles, and full adsorption of protein antigens.

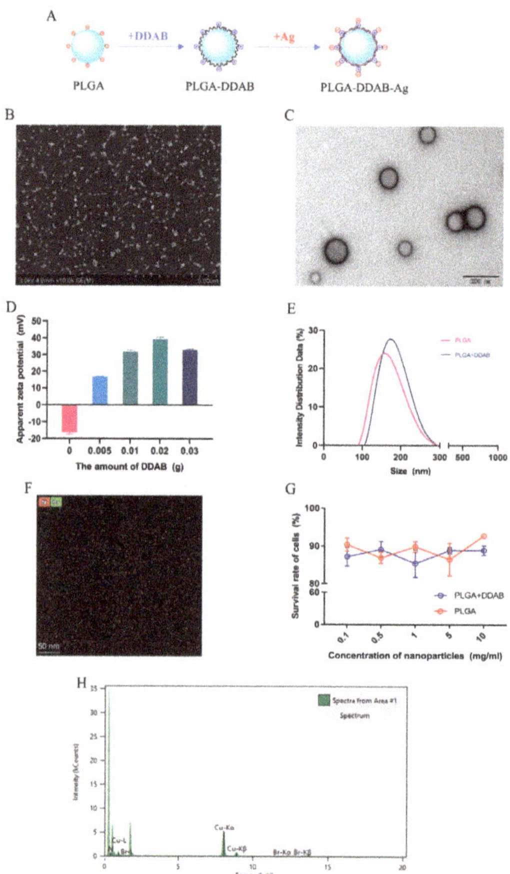


**Figure 2.** Characterization of nanoparticles. (**A**) Schematic diagram of PLGA NPs with the introduction of cationic lipid DDAB and antigen (Ag). (**B**) Scanning electron microscopy image of PLGA-DDAB NPs (scale bar: 5 μm). (**C**) Field emission transmission electron microscopy image of PLGA-DDAB NPs (scale bar: 200 nm). (**D**) Effect of the amount of DDAB on the surface ZETA potential of PLGA NPs. (**E**) Comparison of particle size and distribution between PLGA and PLGA-DDAB NPs. (**F**) Testing images of Br and N elements in PLGA-DDAB NPs. (**G**) Cell viability of MDCK cells after co-incubation with different concentrations of PLGA and PLGA-DDAB NPs. (**H**) EDS spectrum of PLGA-DDAB NPs.

**Table 1.** Introducing DDAB resulted in changes in the mean ± standard error of size, zeta potential, and PDI of PLGA nanoparticles.

| Group | Size (nm) | Zeta Potential (mv) | PDI |
|---|---|---|---|
| PLGA | 164.0 ± 37.4 | −16.2 ± 1.0 | 0.157 ± 0.016 |
| PLGA-DDAB | 176.7 ± 33.4 | 39.3 ± 1.2 | 0.052 ± 0.022 |

We evaluated the adsorption efficiency and safety of the prepared PLGA-DDAB NPs to ensure the nanoparticles could be used in subsequent mouse immunization experiments. The concentration of rETX$^{Y196E}$-C protein in the supernatant before and after adsorption by PLGA-DDAB NPs was measured using the bicinchoninic acid (BCA) assay method and the adsorption efficiency of PLGA-DDAB NPs to rETX$^{Y196E}$-C protein calculated at approximately 60%. The cytotoxicity of the nanoparticles was measured by co-culturing Madin–Darby canine kidney (MDCK) cells with PLGA NPs and PLGA-DDAB NPs. The survival rate of MDCKS cells was approximately 90% for both types of nanoparticles (Figure 2G), indicating good biocompatibility of PLGA-DDAB NPs for subsequent vaccine delivery.

### 2.3. *Analysis of the Performance of PLGA-DDAB dMN*

We used fish gelatin and sucrose as MN matrices and prepared PLGA-DDAB dissolving microneedle patches (dMN) using a mold (Figure 3A). We used a two-step method to prepare the MN tips and backings separately to ensure that the vaccine was concentrated in the tip of the needle. The unmolded MN patch was placed under a stereomicroscope for observation; Figure 3B shows that the 14 × 14 MN array was complete and the tips were intact.

We used PLGA fluorescent NPs encapsulating rhodamine B to verify whether the dMN prepared in this experiment concentrated the drug at the tip of the needle. MNs were made using the same preparation method, and the position of the fluorescent nanoparticles was observed using a fluorescence inverted microscope. The PLGA fluorescent NPs were mainly concentrated at the tip of the needle (Figure 3F), indicating that the antigen was concentrated at the tip of the MN and could achieve accurate and quantitative delivery. The PLGA-DDAB/rETX$^{Y196E}$-C MN was quantified using a sandwich Elisa method, and the average content of rETX$^{Y196E}$-C in the PLGA-DDAB dMN was calculated to be 9.86 ± 2.47 μg/patch, which was equivalent to that of the needle group.

To evaluate the ability of the PLGA-DDAB dMN to penetrate the skin, mechanical performance evaluation was carried out using a force-displacement testing machine. As shown in Figure 3C, after the force gauge probe was moved 0.3 mm towards the backing of the patch with an axial pressure of 20 N, only the tip of the MN was slightly deformed (Figure 3E), indicating that PLGA-DDAB dMN can withstand at least 20 N of pressure and has sufficient mechanical strength to pierce the mouse epidermis layer and deliver the antigen subcutaneously. Optical coherence tomography (OCT) can reveal significant breaks in the epidermal layer of pig skin caused by the penetration of microneedles. As the dMN gradually dissolves within the skin, these breaks also disappear in the OCT images. This phenomenon can reveal the dissolution time of the dMN after piercing the skin. The dMNs had mostly dissolved in pig skin 10–15 min after piercing (Figure 3G), indicating that the dMN had good biocompatibility and could dissolve into the skin in a relatively short time.

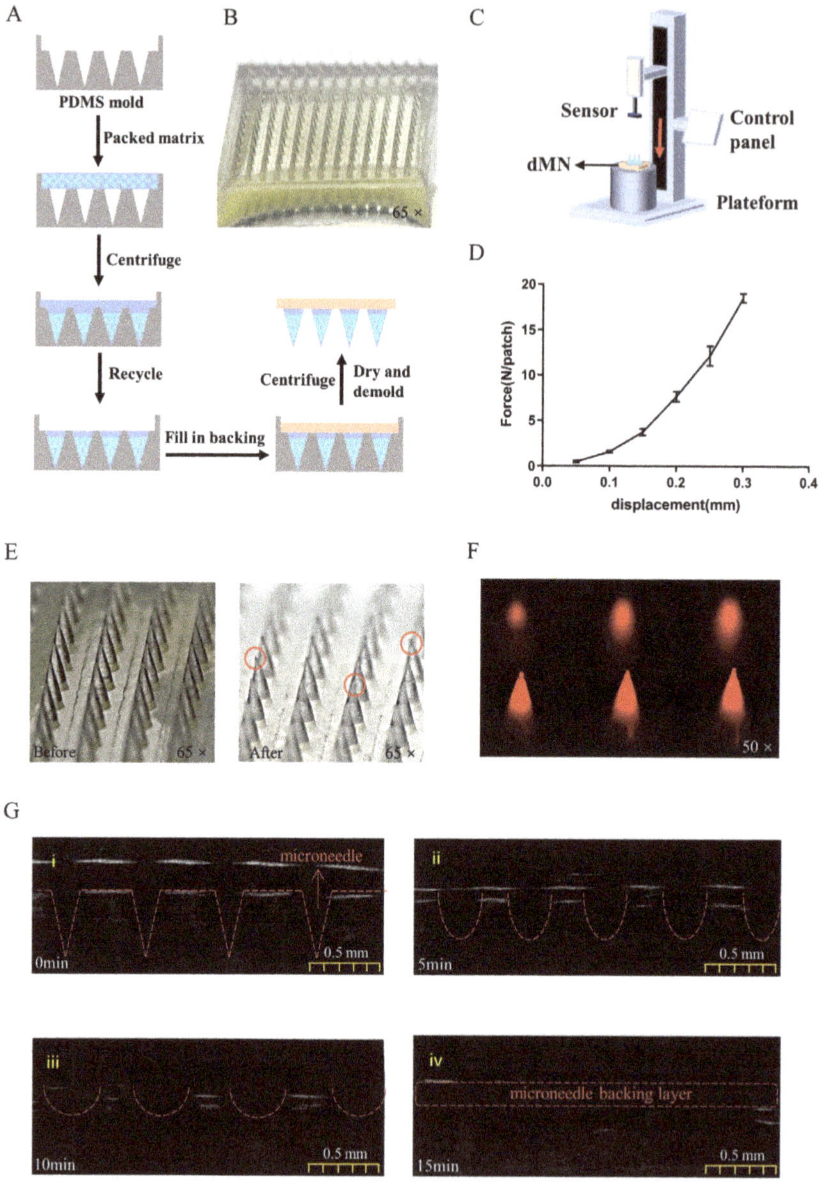


**Figure 3.** Preparation and evaluation of PLGA-DDAB dMNs. (**A**) Schematic diagram of PLGA-DDAB dMN preparation process. (**B**) Body mode microscope image of dMN. (**C**) Schematic diagram of dMN force-displacement testing. (**D**) Force-displacement curve of PLGA-DDAB dMN. (**E**) Microscopic images of dMN before and after compression, with red circles indicating needle tip deformation. (**F**) Fluorescence image of PLGA-DDAB dMN encapsulating Rhodamine B. (**G**) Optical coherence tomography (OCT) image of PLGA-DDAB dMN dissolved after in vitro skin insertion (scale bar is 0.5 mm). (**i**): 0 min, (**ii**): 5 min, (**iii**): 10 min, (**iv**): 15 min images after the microneedle was inserted into the skin.

### 2.4. *In Vivo Release Time and Antibody Titer Analysis of MN Vaccines*

The indirect ELISA method was used to estimate the serum titers of mice immunized with subcutaneous (SC) or dMN immunizations in each of three groups: Ag, Ag + Al, and Ag + PLGA groups. The SC Ag group had a low antibody titer, but the other five groups all produced antibody titers above $10^4$ (Figure 4C). Although the antibody titers of the Ag + PLGA groups were higher than those of the Ag and Ag + Al groups, the difference was not statistically significant. Only the antibody titer of the SC Ag + PLGA group showed a statistically significant difference from that of the SC Ag group after the second and third immunizations. Therefore, we believe that both the Ag + Al and Ag + PLGA groups of SC immunization and dMN immunization can produce strong immune effects.

To analyze whether Al adjuvant and PLGA NPs adjuvant have similar immune-enhancing mechanisms, a mouse in vivo imaging experiment was performed to compare the antigen release time in vivo. The rETX$^{Y196E}$-C protein labeled with Cy5.5 dye was delivered using the same immunization methods as above, and the fluorescent signal was collected using the IVIS Spectrum imaging system (PerkinElmer, Waltham, MA, USA). Interestingly, we found that there was no correspondence between the antigen release time and the antibody titer. Both SC and dMN immunizations with Ag groups showed no fluorescence signal after 2 d post-vaccine (Figure 4A), yet the antibody titer of the dMN Ag group was higher (Figure 4C). This result shows that the use of dMN has the potential to improve immune responses. The dMN Ag + Al group stopped showing an antigen fluorescence signal more than 2 days before the SC group. Similarly, the dMN Ag + PLGA group stopped showing a signal more than 5 d before the SC group (Figure 4A), but the antibody titer between PLGA and AI groups did not differ and was high in both groups (Figure 4C). The prolonged antigen release time of the AI immunization group is indicative of the sustained release effect of the aluminum adjuvant, which enhances the immune response [23]. No sustained release phenomenon was observed in either Ag + PLGA group, and thus, the immune-enhancing mechanisms of the two adjuvants clearly differ. We speculate that the Ag + PLGA immune-enhancing mechanism may be related to its easier uptake by antigen-presenting cells, but further exploration of this mechanism is needed.

### 2.5. *Analysis of Immune Response Types*

To evaluate whether Ag + PLGA can effectively promote Th1 and Th2 immune responses, we estimated the type of immune response induced by different adjuvants by measuring the IgG subtypes (IgG2a and IgG1) in mouse serum and cytokines in splenocytes after immunization. The ratio of IgG2a to IgG1 and cytokines represents the polarization effect of the immune response in both Th1 and Th2 aspects. Unlike aluminum adjuvant, which tends to stimulate Th2 immune response, PLGA NPs generated a higher ratio of IgG2a to IgG1 (Figure 5A) and significantly promoted production of IL-2 and TNF-$\gamma$ (Figure 5D,E), indicating an induced Th1 immune response, and activation of CD4$^+$ T and CD8$^+$ T cells (Figure 5B,C), further demonstrating that Ag + PLGA effectively enhances cellular immunity. In addition, the IL-4 and IL-6 induced by Ag + PLGA were both higher than those induced by the aluminum adjuvant (Figure 5F,G), indicating that they also enhance humoral immunity. Based on these results, dMN delivery can achieve the same immune effect as subcutaneous injection, and Ag + PLGA can effectively enhance immunogenicity by stimulating the production of two types of immune response.

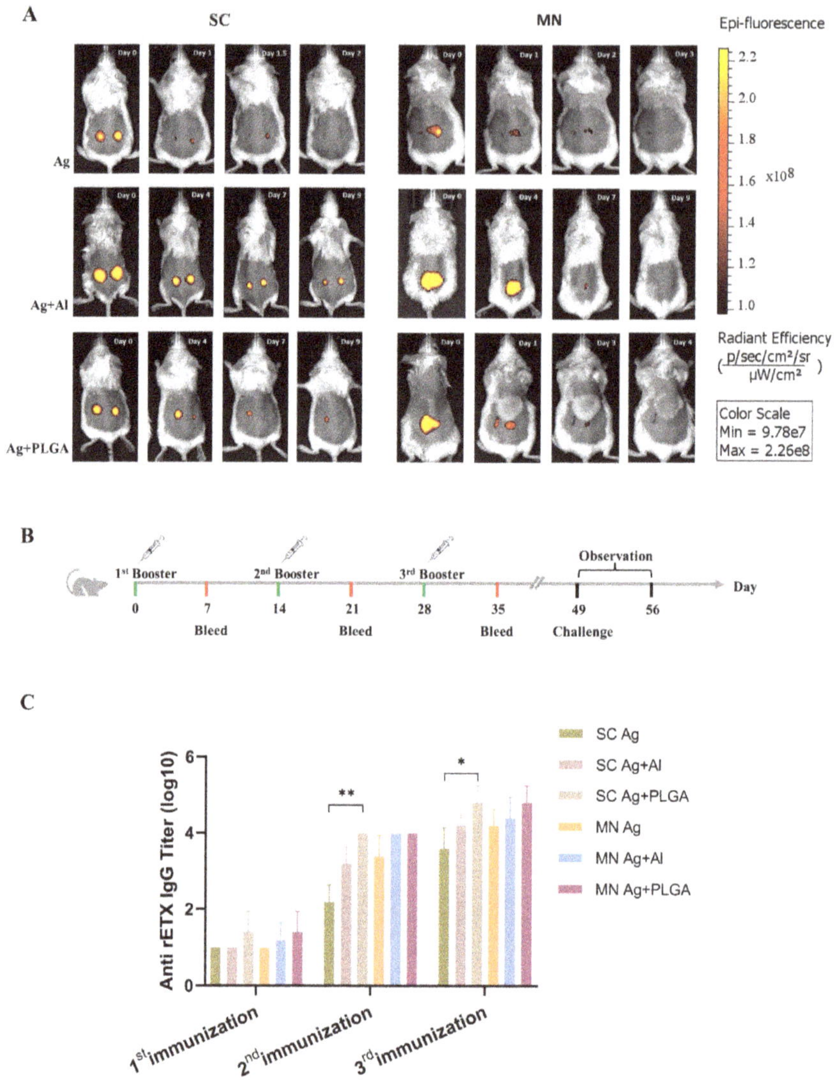


**Figure 4.** (**A**) Comparison of in vivo release time of rETX$^{Y196E}$-C protein in mice. The relative residence times of the antigen at the injection site for the subcutaneous immunization group and the MN group are Ag + Al > Ag + PLGA > Ag. (**B**) Timeline of the immunization process. (**C**) Antibody titers in the serum of mice from the six immunization groups after three immunizations, * $p < 0.05$, ** $p < 0.01$. MN: microneedle patch; SC: subcutaneous injection.

### 2.6. Protective Effect of Ag + PLGA dMN Vaccine

The protective effect of the dMN vaccine on mice was explored through in vivo neutralization experiments, in vitro neutralization experiments, and pathological analysis.

In the in vivo neutralization experiment, immunized mice were intraperitoneally injected with 100 × $LD_{50}$ GST-ETX toxin and observed for 3 d. All mice in the PBS control group died within 6 h; two out of three subcutaneously immunized groups died, including all mice in the SC immunization Ag group within 24 h; one mouse in the SC immunization Ag + Al group died during the 72 h observation period, while all mice in the SC immunization Ag + PLGA group survived (Figure 6A). All three groups of mice immunized with dMN containing the same antigenic components survived (Figure 6A); even the dMN Ag group without any adjuvant was able to resist the attack of 100 × $LD_{50}$ GST-ETX toxin.

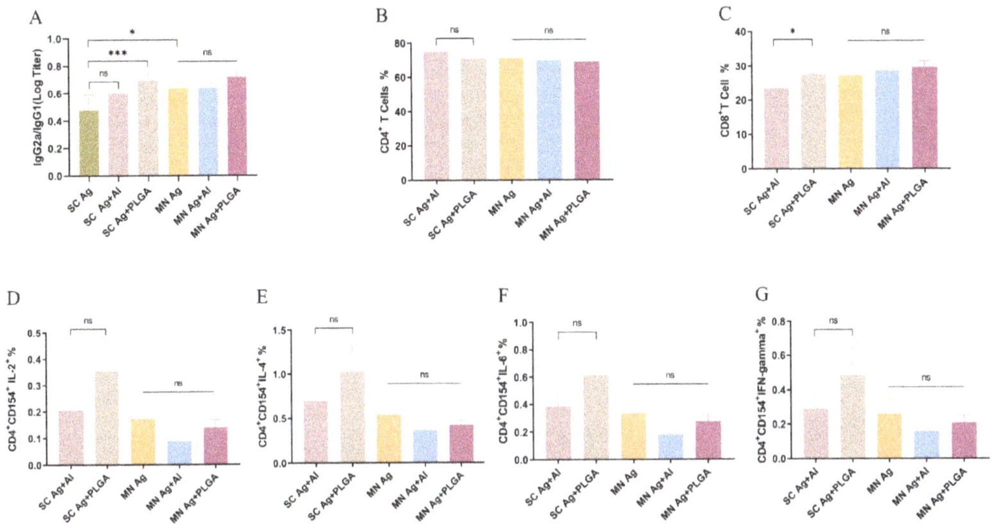

**Figure 5.** (**A**) IgG2a/IgG1 ratio of specific antibodies in mouse serum on day 35. (**B**) Percent CD4$^+$ T cells. (**C**) Percent CD8$^+$ T cells. (**D**) Percent CD4$^+$CD154$^+$ IL-2$^+$. (**E**) Percent CD4$^+$CD154$^+$IL-4$^+$. (**F**) Percent CD4$^+$CD154$^+$IL-6$^+$. (**G**) Percent CD4$^+$CD154$^+$IFN-gamma$^+$. * $p < 0.05$, *** $p < 0.001$. MN: microneedle patch; SC: subcutaneous injection.

The in vitro neutralization experiment included both animal and cell experiments. The immune sera of each of the six groups of mice were mixed with GST-ETX toxin, incubated at 37 °C for 30 min, and then injected into blank mice and co-cultured with cells to observe the survival of mice and cells. Both mouse survival (Figure 6B) and cell proliferation curves (Figure 6D) showed that the immune sera of the SC immunization Ag group did not protect either mice or cells, and the immune sera of the SC immunization Ag + Al group protected mice but did not effectively protect cells. The immune sera of the SC immunization Ag + PLGA group and the three groups of mice immunized with dMN were able to protect naive mice from 10 × $LD_{50}$ GST-ETX toxin after a 20-fold dilution and protect cells from the attack of 20 × $CT_{50}$ GST-ETX toxin (see Figure 6C). These results are consistent with those of the in vivo neutralization experiment.

For pathological analysis, we used HE staining to observe the kidney and brain of mice at death or 7 d after toxin injection. Positive control mice were intraperitoneally injected with 100 × $LD_{50}$; intraperitoneal injection of PBS was used as a negative control. Positive control mice lost mobility within 10 min of injection and were euthanized immediately when they were close to death. Pathological section results (Figure 6E) showed that the positive control mice had significant bleeding and edema in the kidney, and significant vacuolization, bleeding, and edema in the brain, while there were no pathological changes in the kidney and brain tissue of the negative control and immunized mice, except for the SC immunization Ag group. These results indicate that all immunization groups except for the SC immunization Ag group were protected from the toxin.

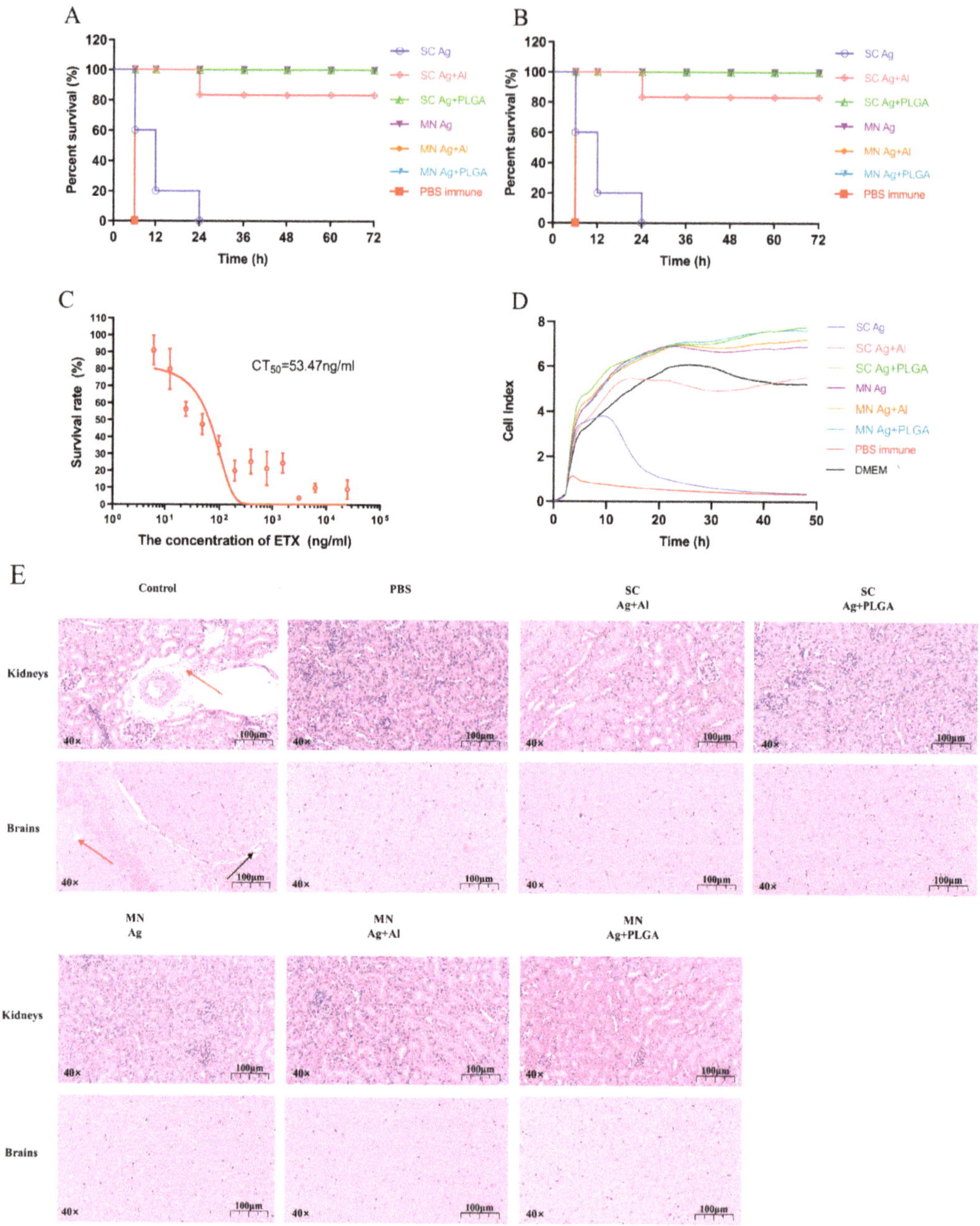


**Figure 6.** (**A**) Survival curve of mice after toxin challenge. (**B**) In vitro neutralization assay of animal toxicity. (**C**) $CT_{50}$ of GST-ETX on MDCK cells. (**D**) In vitro neutralization assay of cell toxicity: cell proliferation curve. (**E**) Pathology of mouse kidney and brain after toxin challenge. The positive control group exhibited kidney bleeding and edema (red arrow), brain tissue vacuolization (red arrow), and edema (black arrow). No pathological injury was observed in any of the immunized groups or the PBS negative control group. MN: microneedle patch; SC: subcutaneous injection.

## 3. Discussion

A dMN containing PLGA NPs was successfully prepared for the delivery of the $rETX^{Y196E}$-C protein vaccine in this study. The patch has advantages, such as enhanced immune response, quantitative delivery, and significantly reduced pain, and effectively protected mice against the toxin at $100 \times LD_{50}$ GST-ETX.

PLGA is an FDA-approved biodegradable polymer with good biocompatibility, and microspheres prepared using PLGA as an adjuvant or delivery system have gained wide attention. PLGA can interact with antigen-presenting cells to induce an ideal immune response and has been used as an adjuvant in vaccines against various diseases, including hepatitis B [9,10], malaria [11,12], and tumors [24]. PLGA has also been shown to increase antibody titers to some extent. However, there are also certain problems in the application process, such as the complex and cumbersome coupling process between the antigen and PLGA NPs, which can lead to antigen inactivation, the toxicity of crosslinking or coupling agents that render the vaccine unusable, and prolonged antigen release from PLGA NPs leading to immune tolerance. In our study, cationic lipid DDAB was introduced into the PLGA NPs we prepared to give them a positive surface charge, allowing a gentle and efficient loading of the mutant protein vaccine $rETX^{Y196E}$ -C prepared in our laboratory, reducing the impact of the coupling process on antigen activity.

In the evaluation of immune effects, we found that, although both aluminum adjuvant and PLGA NPs can stimulate mice to increase antibody titers, their mechanisms of enhancement differ, with aluminum adjuvant having a longer release time. Although the PLGA NPs' release time is somewhat extended, it is significantly weaker than for the aluminum adjuvant. Therefore, we speculate that the immune enhancement mechanism of PLGA NPs may be related to their easier uptake by antigen-presenting cells. Immunoassay results indicate that PLGA NP can be used as an alternative to aluminum adjuvant. In the neutralization experiments, the protective effect of dMN immunization on mice was superior to subcutaneous immunization, consistent with previous reports [25–27], showing that MN immunization also has an immune enhancement effect. Therefore, the combined use of PLGA NPs and dMN immunization can effectively enhance the immune effect of vaccine antigens.

The issue of accurate quantitative delivery of vaccine proteins in dMN remains a challenge, impeding the application of such vaccines. Various research teams are attempting to solve this problem. For example, the Prausnitz team has used different polymers to prepare the tip and backing of the MN, concentrating the drug into the tip and thereby avoiding diffusion to the backing [28]. They also employed freeze-drying to prepare the tip for quick separation from the backing, enabling complete delivery of the vaccine antigen into the dermis. Several Chinese teams, including Xin Dong Guo and Chenjie Xu, have improved the preparation process to form a bubble cavity between the tip and the backing, facilitating complete drug delivery into the skin [29–32]. In this study, we employed centrifugation during the MN preparation process to concentrate the PLGA NPs adsorbed with vaccine proteins at the tip of the MN, thereby preventing diffusion to the backing and achieving quantitative delivery of vaccine antigens to the dermis.

This dMN vaccine patch has the benefits of quick and easy administration, painlessness, and excellent thermal stability, which are characteristic of dMN vaccine patches. In addition, it is enhanced with PLGA NPs to provide a more robust immune response. Furthermore, the preparation method is simple, safe, effective, and easily scalable for mass production. This patch provides an effective means of preventing ETX enterotoxemia, reducing losses in animal husbandry resulting from ETX toxicity, and guarding against the biological security threat posed by ETX as a bioterrorism agent.

## 4. Conclusions

We have successfully prepared PLGA NPs with a smooth surface, uniform size, good safety, and positive charge that can gently and efficiently adsorb $rETX^{Y196E}$-C vaccine antigens. These nanoparticles are combined with dMN to deliver antigens. We validated the reliability of the Ag + PLGA dMN vaccine patch in terms of mechanical properties, skin penetration, immunogenicity, and protection. We show that the patch had excellent mechanical properties and can withstand a pressure of 20 N without significant deformation while successfully penetrating the skin and achieving transdermal vaccine delivery. The vaccine enhanced both cellular and humoral immune responses, with a specific IgG antibody titer of $10^5$ after three immunizations, and effectively activated T cells and the expression of cytokines such as IL-2, IL-4, IL-6, and interferon-gamma. This approach provides a new powerful tool for preventing ETX intoxication and bioterrorism attacks and has the potential to become a new vaccine delivery system for other vaccines or drugs.

## 5. Materials and Methods

### 5.1. Animals

SPF-grade female Balb/c mice (6–8 weeks old) were purchased from SPF (Beijing, China) Biotechnology Co., Ltd. The experimental mice were housed in the SPF animal room of the Experimental Animal Center of the Academy of Military Medical Sciences, and the animal experiments were reviewed and approved by the Animal Ethics Committee of the Academy of Military Medical Sciences. All animal experiments complied with the requirements of the Guidelines for Ethical Review of Laboratory Animal Welfare of the Academy of Military Medical Sciences.

### 5.2. Preparation of Vaccine Antigens and Toxin Proteins

The rETX$^{Y196E}$-C and GST-ETX recombinant strains were stored in our laboratory, and detailed protein preparation methods followed that of previous publications [32–34]. Briefly, the bacterial strains were resuscitated in LB medium and, after expansion culture, the bacterial cells were disrupted using ultrasonic treatment. The supernatant containing the target protein was obtained by centrifugation at 11,000× *g* for 20 min. Nickel-nitrilotriacetic (GE Healthcare, Piscataway, NJ, USA) and glutathione Sepharose (GE Healthcare, Piscataway, NJ, USA) were used to perform affinity chromatography on the rETX$^{Y196E}$-C and GST-ETX recombinant proteins, respectively. The purified products were concentrated using a 10 kDa ultrafiltration tube, and imidazole in the solution was displaced from the system by repeating the process three times. The molecular weight and purity of the purified products were confirmed by SDS-PAGE gel electrophoresis. A BCA protein quantitation kit (PA101-01, Biomed, Beijing, China) was used to determine protein concentration.

### 5.3. Preparation and Characterization of PLGA-DDAB Nanoparticles

PLGA-DDAB NPs were prepared using an emulsion evaporation method [6,35]. Briefly, PLGA (50:50 poly (DL-Lactide-co-glycolide), carboxylate end group (nominal); LACTEL Absorbable Polymers, Birmingham, AL, USA) and DDAB (Sigma-Aldrich, St. Louis, MO, USA) were dissolved in 10 mL of DCM (Sinopharm Chemical Reagent Co., Ltd., Shanghai, China) was used as the oil phase and 0.5 g of PVA (Sigma-Aldrich, St. Louis, MO, USA) was dissolved in 50 mL of dd$H_2O$ as the aqueous phase. The oil phase was slowly added to the aqueous phase at 600 rpm and emulsified for 30 min (125 W, 75%) using an ultrasonic disintegrator (Q125; Qsonica LLC, Newton, CT, USA). The large pellets were precipitated by centrifugation at 4000× *g* for 10 min, the supernatant was resuspended in dd$H_2O$ and washed three times by centrifugation at 20,000× *g* for 30 min to obtain PLGA-DDAB NPs, which were resuspended in 2 mL of dd$H_2O$ with 3% sucrose (V900116, Sigma, St. Louis, MO, USA) as lyophilization protectant.

The same method was used for the preparation of fluorescent PLGA-DDAB NPs. Rhodamine B was dissolved in dd $H_2O$ as the internal aqueous phase, PLGA and DDAB were dissolved in DCM as the oil phase, and the internal aqueous phase was added to the oil phase in a ratio of 1:5 internal aqueous phase:oil phase, sonicated for 2 min (125 W, 75%), and then added to 1% PVA (*w*/*v*) external aqueous phase solution and sonicated for 30 min (125 W, 75%). Water-in-oil-in-water emulsions were obtained; the organic solvent was evaporated by stirring overnight in a fume hood protected from light, $ddH_2O$-washed three times and then lyophilized to obtain fluorescent PLGA-DDAB NPs.

PLGA-DDAB NP surface morphology was observed using a field emission scanning electron microscope (SU8020, Hitachi Limited, Tokyo, Japan) and a field emission transmission electron microscope (FEI Talos F200X, FEI Company, Hillsboro, OR, USA). The particle size distribution, dispersibility index (PDI), and ZETA potential were determined using a laser particle sizer (Zetasizer Nano ZS90, Malvern Instruments Ltd., Malvern, UK). Energy dispersive X-ray spectroscopy (EDX) was used to analyze the presence of Br and N elements in the DDAB within the PLGA nanoparticles.

### 5.4. Adsorption Rate Testing

To detect the adsorption rate of $rETX^{Y196E}$-C protein onto PLGA-DDAB NPs, the following series of steps was performed. PLGA-DDAB NPs in freeze-dried powder form were reconstituted in 500 μL of ultrapure water. The mixture was then sonicated to achieve dispersion. A predetermined volume of $rETX^{Y196E}$-C protein solution at a specific concentration was added to the PLGA-DDAB NPs dispersion. The solution was then gently mixed at 4 °C for a defined period to allow for the formation of PLGA-DDAB/$rETX^{Y196E}$-C nanoparticle suspension. Subsequently, the suspension was centrifuged at 20,000× *g* for 3 min and the supernatant was collected. The concentration of $rETX^{Y196E}$-C protein in the supernatant was estimated using the bicinchoninic acid (BCA) assay method. The amount of $rETX^{Y196E}$-C protein adsorbed onto PLGA-DDAB nanoparticles was calculated by subtracting the concentration of $rETX^{Y196E}$-C protein in the supernatant after adsorption from the initial concentration of added antigen.

### 5.5. Cell Viability Assay for PLGA-DDAB Nanoparticles

MDCK cells were seeded in 96-well plates at a density of $1 \times 10^5$ cells/mL and incubated overnight at 37 °C with 5% $CO_2$. PLGA-DDAB NPs were diluted in DMEM containing 10% FBS at different concentrations and added to 96-well plates. The plates were incubated for 1 h at 37 °C with 5% $CO_2$, then the medium was removed and the cells were washed with PBS three times. Fresh medium (100 μL) and MTS (20 μL) were added to each well and the plates were incubated under the same conditions for 3 h. The absorbance at 492 nm was measured using a microplate reader (Multiskan Mk3, Thermo Fisher, Waltham, MA, USA), and cell viability was calculated accordingly.

### 5.6. Preparation and Characterization of PLGA-DDAB dMN

The PLGA-DDAB dMN was prepared using a two-step method and a PDMS mold provided by Shanghai Jiao Tong University [36]. The MNs were conical in shape with a height of 650 μm, a bottom diameter of 360 μm, and a spacing of 360 μm [37]. In the first step, the needle tip of the MN was prepared. PLGA-DDAB containing $rETX^{Y196E}$-C protein or blank PLGA-DDAB was suspended in a solution of 35% gelatin (G7041, Sigma, USA) (m/v) and 10% sucrose (m/v) in water. The solution was centrifuged at 3500 rpm for 10 min (Beckman Coulter, 21R Centrifuge, San Pablo, CA, USA) to allow the solution to completely enter the needle tip. Any excess solution was then removed. In the second step, the backing of the MN was prepared. A water solution of the same matrix was added, and the mold was centrifuged at 3500 rpm for 10 min. The mold was then removed and placed in a drying oven for 24 h. The MN patch was carefully removed with forceps and sealed for storage.

### 5.7. MN Performance Assessment

Mechanical properties of the PLGA-DDAB dMN was evaluated using a force-displacement testing machine (MARK-10, Copiague, NY, USA). The MNs were placed horizontally on the platform of the testing machine with the needle tips facing upward. The force gauge probe was set to move downwards at a uniform speed of 0.1 mm/s until the maximum force of 20 N·m was reached. The force-displacement curves were analyzed, and the needle tip deformation was observed using an optical microscope (Olympus DP71, Tokyo, Japan) after force application.

The PLGA-DDAB dMN was tested for skin penetration using pig skin and optical coherence tomography (MDL, VivoSight DX, Kent, UK). The MNs were placed on the pig skin, and pressure was applied using the thumb for 0, 5, 10, and 15 min. The patches were then left on the skin, and OCT scanning was performed immediately to visualize the skin penetration of the MN.

### 5.8. *In Vivo Release Time Determination*

The experimental mice were divided into two immunization (SC, MN) by three vaccine-type (Ag, Ag + Al, Ag + PLGA) groups for six groups total: SC immunization with antigen group, SC immunization with Ag + Al group, SC immunization with Ag + PLGA group, MN Ag group, MN Ag + Al group, and MN Ag + PLGA group. Prior to administration, the back hair of mice was removed with depilatory cream and then wiped with saline and dried with a dry cotton ball. The doses of each group were kept the same, and the subcutaneous needle injection group was injected subcutaneously at two points on the back of the mouse with a total volume of 200 µL. In the MN group, the patch was pressed against the skin on the back of the mouse with fingers for 10–15 min. After removal of the patch, any remaining liquid on the skin surface was wiped off with a cotton ball. Fluorescence signals were collected using an IVIS Spectrum (PerkinElmer, Waltham, MA, USA) with excitation light set at 640 nm and emission light at 700 nm. The exposure time was 5 s, and fluorescence changes were observed.

### 5.9. *PLGA-DDAB-rETX$^{Y196E}$-C dMN Immunization in Mice*

SPF-grade female Balb/c mice aged 6–8 weeks were obtained from Beijing Sibeifu (Beijing, China) Biotechnology Co., Ltd. The mice were randomly divided into eight groups, with five mice in each group (see Table 2 for details). The immunization schedule is shown in Figure 5A, with immunizations administered on days 0, 14, and 28. The immunization dose and route were consistent for each immunization. Blood samples were collected from the tail vein of each mouse one week after each immunization to measure antibody titers.

**Table 2.** Vaccination experiment groups.

| No. | Group ($n$ = 5) | | Dose of Antigen (µg) | NPs Content (µg) |
|---|---|---|---|---|
| 1 | MN | Ag | 10 | - |
| 2 | | Ag + Al | 10 | - |
| 3 | | Ag + PLGA | 10 | 400 |
| 4 | | PBS | - | - |
| 5 | SC | Ag | 10 | - |
| 6 | | Ag + Al | 10 | - |
| 7 | | Ag + PLGA | 10 | 400 |
| 8 | | PBS | - | - |

MN: microneedle patch; SC: subcutaneous injection.

### 5.10. *Mouse Challenge Experiment*

Natural ETX toxin challenge experiments were performed 14 days after the third immunization in mice. Mice were injected with 500 µL of 100 × $LD_{50}$ of toxin via intraperitoneal injection. The survival status of each group of mice was recorded for 3 consecutive days.

### 5.11. *ELISA Detection of Antibody Levels in Immune Sera and the Content of MN Vaccine*

The titers of ETX-specific antibodies IgG (Solarbio C1055, Beijing, China), IgG1 (ab97240, Abcam, UK), and IgG2a (ab97245, Abcam, UK) were measured by indirect ELISA. In brief, rETX$^{Y196E}$-C protein was diluted to 10 µg/mL with Elisa coating buffer (diluted to 1×, Solarbio C1055, Beijing, China) and added to 96-well ELISA plates (Thermo

442404, Waltham, MA, USA) at 100 μL per well, and incubated overnight at 4 °C. After washing with PBST three times, 5% BSA was added and the plate incubated at 37 °C for 1 h. After washing with PBST three times, mouse sera were diluted with 5% BSA in concentration gradients of $10^{-1}$, $10^{-2}$, $10^{-3}$, $10^{-4}$, $10^{-5}$, and $10^{-6}$, added to ELISA plates at 100 μL per well, and incubated at 37 °C for 1 h. After washing with PBST three times, horseradish peroxidase-labeled goat anti-mouse secondary antibodies IgG (1:5000 dilution), IgG1 (1:10,000 dilution), and IgG2a (1:10,000 dilution) were added to each well and wells were incubated at 37 °C for 1 h. After washing with PBST three times, TMB colorimetric substrate (Solarbio PR1210, Beijing, China) was added at 100 μL per well, and plates were then incubated for 10 min in the dark. Finally, the reaction was stopped by adding 50 μL of ELISA stop solution (Solarbio C1058, Beijing, China) to each well, and the $OD_{450}$ values were immediately measured. The obtained OD values were corrected for the background interference using the blank control OD value, and immune serum samples with OD values greater than 2.1 times the negative control OD value were considered positive. The MN rETX$^{Y196E}$-C recombinant protein content was determined using the sandwich ELISA method established in our laboratory, detailed as described in reference [38].

### 5.12. In Vitro Neutralization Assay

In vitro neutralization animal toxicity assay. Female Balb/c mice at 6–8 weeks of age were selected for the experiment, with five mice in each group. Mouse sera were obtained after triple immunization and diluted 20-fold. Equal volumes of each diluted serum and $10 \times LD_{50}$ GST-ETX were mixed and incubated at 37 °C for 30 min. The mixture was then intraperitoneally injected into each mouse at a volume of 500 μL and the survival status of the mice in each group was observed.

In vitro neutralization cytotoxicity assay. The $CT_{50}$ of GST-ETX on MDCK cells was measured using the MTS method. E-plate wells were filled with 50 μL of DMEM culture medium containing 10% FBS and placed on the detection platform of the instrument to measure the baseline for 50 min. The MDCK cells with good growth status were resuspended with DMEM culture medium and counted. The cell concentration was adjusted to $2 \times 10^6$ mL$^{-1}$ with DMEM culture medium and plated in the E-plate at 50 μL per well. Mouse serum obtained from a mouse in the immunized group was diluted 20-fold, and GST-ETX was diluted to $20 \times CT_{50}$, with equal volumes mixed and incubated at 37 °C for 30 min. Next, 150 μL of the mixture was added to each well, and duplicate wells were prepared for each concentration. The E-plate was placed on the detection platform, and after incubating at 37 °C for 10 min, the instrument was run. After 72 h, the E-plate was removed and the data were analyzed using the RTCA Software 2.1.5.

### 5.13. Cytokine Detection of Splenocytes Stimulated by rETX$^{Y196E}$-C In Vitro

Splenocyte cytokine flow cytometric analysis was performed on mice that were challenged with the toxin for 7 days, with three mice in each group. Immediately after cervical dislocation, the spleen was removed and minced in a biosafety cabinet, and then gently ground and suspended in 1640 medium containing 10% FBS, 0.1 mM 2-mercaptoethanol, 100 U/mL penicillin, and 100 U/mL streptomycin to make a single cell suspension. Four million cells were plated in a 12-well plate, and to each well was added 2 μg anti-mouse CD28 antibody (553294, BD Biosciences, San Jose, CA, USA), 2 μg anti-mouse CD49d antibody (553153, BD Biosciences, USA), and 40 μg rETX$^{Y196E}$-C (positive control added 20 ng anti-mouse CD3 antibody (100339, BD Biosciences, USA), negative control added only 2 μg anti-mouse CD28 antibody and 2 μg anti-mouse CD49d antibody), as well as 1 μg CD154 BV605 (745242, MR1, BD Biosciences, USA) and 1 μg CD107a BV786 (564349, 1D4B, BD Biosciences, USA), and the appropriate amount of culture medium was added to a total volume of 2 mL. The cell culture plate was placed in a $CO_2$ incubator at 37 °C with 5% $CO_2$ for 18 h. Two microliters of BFA (555029, BD Biosciences, USA) and 1.4 μL of monensin (554724, BD Biosciences, USA) were added per milliliter of medium, and the cells were further incubated at 37 °C and 5% $CO_2$ for another 6 h. The cells were transferred to flow tubes (352054, BD Falcon, USA), and the following reagents were added in sequence: receptor blocking reagent (553141, BD Biosciences, USA); surface staining antibody mix (CD3 PerCP-eFluor710 (46-0032-82, 17A2, Invitrogen, Waltham, MA, USA); CD4 R718 (553046, RM4-5, BD, USA); CD8 BV510 (100752, 53-6.7, Biolegend, San Diego, CA, USA); wash buffer (DPBS containing 1% FBS); live/dead dye (423105, Biolegend, USA); fixation solution; permeabilization washing buffer (554714, BD Biosciences, USA); intracellular staining antibody mix (IL-2 APC (503810, JES6-5H4, Biolegend, USA); IL4 BV421 (504120, 11B11, Biolegend, USA); IL-6 PE (554401, MP5-20F3, BD Biosciences, USA); IFN-gamma Alexa Fluor 488 (505813, XMG1.2, Biolegend, USA); wash buffer; and PBS. Data were analyzed using flow cytometry (Aurora, Cytek, Fremont, CA, USA) and SpectroFlo software (Cytek, V2.2.0.4).

### 5.14. Pathological Damage in rETX$^{Y196E}$-C Immunized Mice

The brain and kidneys of mice were taken to assess pathological damage after the toxin challenge. The positive control group mice were injected intraperitoneally with $100 \times LD_{50}$ GST-ETX, and the negative control group consisted of newly purchased mice. The brain and kidneys of the mice were taken under sterile conditions, the adipose tissue carefully removed, and then washed with PBS to remove blood. Organs were then fixed in 4% paraformaldehyde (P1110, Solarbio, Beijing, China) for 24 h, embedded, stained with hematoxylin and eosin (HE), and observed using slicing.

### 5.15. Statistical Analysis

GraphPad Prism 8.3.0 was used for statistical analysis. All experiments were repeated at least three times, and data are expressed as means $\pm$ standard deviation. Mouse sera antibody titers were analyzed with a two-way ANOVA; statistically significant differences were defined as $p < 0.05$.

**Author Contributions:** Conceptualization, H.Y. and S.G.; data curation, W.W. and Y.L.; formal analysis, W.W. and Y.L.; funding acquisition, W.X.; methodology, Z.J. and J.W. (Junhong Wang).; resources, W.X.; supervision, L.K.; visualization, X.L. and Y.C.; writing—original draft, W.W. and Y.L.; writing—review and editing, J.W. (Jing Wang) and J.W. (Jinglin Wang). All authors have read and agreed to the published version of the manuscript.

**Funding:** This research was funded by Beijing Nova Program, grant number: Z201100006820028.

**Institutional Review Board Statement:** Ethics approval and consent to participate: All animal work was conducted in strict accordance with recommendations in the Guide for the Care and Use of Laboratory Animals and the American Veterinary Medical Association (AVMA) Guidelines for the Euthanasia of Animals. The protocol was approved by the Institutional Animal Care and Use Committee at the Academy of Military Medical Sciences (AMMS) (protocol code: IACUC-DWZX-2021-053, date of approval: 9 September 2021). All efforts were made to minimize mouse suffering.

**Informed Consent Statement:** Not applicable.

**Data Availability Statement:** The data presented in this study are available on request from the corresponding author.

**Conflicts of Interest:** The authors declare no conflict of interest. The funders had no role in the design of the study; in the collection, analyses, or interpretation of data; in the writing of the manuscript; or in the decision to publish the results.

## References

1. Lewis, M.; Weaver, C.D.; McClain, M.S. Identification of Small Molecule Inhibitors of Clostridium perfringens ε-Toxin Cytotoxicity Using a Cell-Based High-Throughput Screen. *Toxins* **2010**, *2*, 1825–1847. [CrossRef] [PubMed]
2. Lee, J.; Kumar, S.A.; Jhan, Y.Y.; Bishop, C.J. Engineering DNA vaccines against infectious diseases. *Acta Biomater.* **2018**, *80*, 31–47. [PubMed]
3. Shah, R.R.; Hassett, K.J.; Brito, L.A. Overview of Vaccine Adjuvants: Introduction, History, and Current Status. *Methods Mol. Biol.* **2017**, *1494*, 1–13. [CrossRef] [PubMed]

4. Hu, Y.; Smith, D.; Zhao, Z.; Harmon, T.; Pentel, P.R.; Ehrich, M.; Zhang, C. Alum as an adjuvant for nanoparticle based vaccines: A case study with a hybrid nanoparticle-based nicotine vaccine. *Nanomedicine* **2019**, *20*, 102023. [CrossRef] [PubMed]
5. Zhang, W.; Wang, L.; Liu, Y.; Chen, X.; Li, J.; Yang, T.; An, W.; Ma, X.; Pan, R.; Ma, G. Comparison of PLA microparticles and alum as adjuvants for H5N1 influenza split vaccine: Adjuvanticity evaluation and preliminary action mode analysis. *Pharm. Res.* **2014**, *31*, 1015–1031. [CrossRef]
6. Danhier, F.; Ansorena, E.; Silva, J.M.; Coco, R.; Le Breton, A.; Préat, V. PLGA-based nanoparticles: An overview of biomedical applications. *J. Control. Release* **2012**, *161*, 505–522. [CrossRef]
7. Han, S.; Ma, W.; Jiang, D.; Sutherlin, L.; Zhang, J.; Lu, Y.; Huo, N.; Chen, Z.; Engle, J.W.; Wang, Y.; et al. Intracellular signaling pathway in dendritic cells and antigen transport pathway in vivo mediated by an OVA@DDAB/PLGA nano-vaccine. *J. Nanobiotechnol.* **2021**, *19*, 394. [CrossRef]
8. Mönkäre, J.; Pontier, M.; van Kampen, E.E.M.; Du, G.; Leone, M.; Romeijn, S.; Nejadnik, M.R.; O'Mahony, C.; Slütter, B.; Jiskoot, W.; et al. Development of PLGA nanoparticle loaded dissolving microneedles and comparison with hollow microneedles in intradermal vaccine delivery. *Eur. J. Pharm. Biopharm.* **2018**, *129*, 111–121. [CrossRef]
9. Thomas, C.; Rawat, A.; Hope-Weeks, L.; Ahsan, F. Aerosolized PLA and PLGA nanoparticles enhance humoral, mucosal and cytokine responses to hepatitis B vaccine. *Mol. Pharm.* **2011**, *8*, 405–415. [CrossRef]
10. Zheng, X.; Zhu, J.; Zheng, C.; Tan, Z.; Ji, Z.; Tao, J.; Zhao, Y.; Ji, Z.; Hu, Y. Dissolving Microneedle Arrays as a Hepatitis B Vaccine Delivery System Adjuvanted by APC-Targeted Poly (Lactic-co-Glycolic Acid) (PLGA) Nanoparticles. *AAPS PharmSciTech* **2023**, *24*, 42. [CrossRef]
11. Rosas, J.E.; Hernández, R.M.; Gascón, A.R.; Igartua, M.; Guzman, F.; Patarroyo, M.E.; Pedraz, J.L. Biodegradable PLGA microspheres as a delivery system for malaria synthetic peptide SPf66. *Vaccine* **2001**, *19*, 4445–4451. [CrossRef]
12. Carcaboso, A.M.; Hernández, R.M.; Igartua, M.; Rosas, J.E.; Patarroyo, M.E.; Pedraz, J.L. Potent, long lasting systemic antibody levels and mixed Th1/Th2 immune response after nasal immunization with malaria antigen loaded PLGA microparticles. *Vaccine* **2004**, *22*, 1423–1432. [CrossRef]
13. Fairley, S.J.; Singh, S.R.; Yilma, A.N.; Waffo, A.B.; Subbarayan, P.; Dixit, S.; Taha, M.A.; Cambridge, C.D.; Dennis, V.A. Chlamydia trachomatis recombinant MOMP encapsulated in PLGA nanoparticles triggers primarily T helper 1 cellular and antibody immune responses in mice: A desirable candidate nanovaccine. *Int. J. Nanomed.* **2013**, *8*, 2085–2099. [CrossRef]
14. Taha, M.A.; Singh, S.R.; Dennis, V.A. Biodegradable PLGA85/15 nanoparticles as a delivery vehicle for Chlamydia trachomatis recombinant MOMP-187 peptide. *Nanotechnology* **2012**, *23*, 325101. [CrossRef] [PubMed]
15. Boopathy, A.V.; Mandal, A.; Kulp, D.W.; Menis, S.; Bennett, N.R.; Watkins, H.C.; Wang, W.; Martin, J.T.; Thai, N.T.; He, Y.; et al. Enhancing humoral immunity via sustained-release implantable microneedle patch vaccination. *Proc. Natl. Acad. Sci. USA* **2019**, *116*, 16473–16478. [CrossRef] [PubMed]
16. Luo, Z.; Sun, W.; Fang, J.; Lee, K.; Li, S.; Gu, Z.; Dokmeci, M.R.; Khademhosseini, A. Biodegradable Gelatin Methacryloyl Microneedles for Transdermal Drug Delivery. *Adv. Healthc. Mater.* **2019**, *8*, e1801054. [CrossRef]
17. Chu, L.Y.; Ye, L.; Dong, K.; Compans, R.W.; Yang, C.; Prausnitz, M.R. Enhanced Stability of Inactivated Influenza Vaccine Encapsulated in Dissolving Microneedle Patches. *Pharm. Res.* **2016**, *33*, 868–878. [CrossRef]
18. Edens, C.; Collins, M.L.; Goodson, J.L.; Rota, P.A.; Prausnitz, M.R. A microneedle patch containing measles vaccine is immunogenic in non-human primates. *Vaccine* **2015**, *33*, 4712–4718. [CrossRef]
19. Kolluru, C.; Gomaa, Y.; Prausnitz, M.R. Development of a thermostable microneedle patch for polio vaccination. *Drug Deliv. Transl. Res.* **2019**, *9*, 192–203. [CrossRef]
20. Larrañeta, E.; McCrudden, M.T.; Courtenay, A.J.; Donnelly, R.F. Microneedles: A New Frontier in Nanomedicine Delivery. *Pharm. Res.* **2016**, *33*, 1055–1073. [CrossRef]
21. Braz Gomes, K.; D'Souza, B.; Vijayanand, S.; Menon, I.; D'Souza, M.J. A dual-delivery platform for vaccination using antigen-loaded nanoparticles in dissolving microneedles. *Int. J. Pharm.* **2022**, *613*, 121393. [CrossRef] [PubMed]
22. Chen, X.; Liu, Y.; Wang, L.; Liu, Y.; Zhang, W.; Fan, B.; Ma, X.; Yuan, Q.; Ma, G.; Su, Z. Enhanced humoral and cell-mediated immune responses generated by cationic polymer-coated PLA microspheres with adsorbed HBsAg. *Mol. Pharm.* **2014**, *11*, 1772–1784. [CrossRef]
23. Maughan, C.N.; Preston, S.G.; Williams, G.R. Particulate inorganic adjuvants: Recent developments and future outlook. *J. Pharm. Pharmacol.* **2015**, *67*, 426–449. [CrossRef] [PubMed]
24. Koerner, J.; Horvath, D.; Herrmann, V.L.; MacKerracher, A.; Gander, B.; Yagita, H.; Rohayem, J.; Groettrup, M. PLGA-particle vaccine carrying TLR3/RIG-I ligand Riboxxim synergizes with immune checkpoint blockade for effective anti-cancer immunotherapy. *Nat. Commun.* **2021**, *12*, 2935. [CrossRef] [PubMed]
25. Vassilieva, E.V.; Kalluri, H.; McAllister, D.; Taherbhai, M.T.; Esser, E.S.; Pewin, W.P.; Pulit-Penaloza, J.A.; Prausnitz, M.R.; Compans, R.W.; Skountzou, I. Improved immunogenicity of individual influenza vaccine components delivered with a novel dissolving microneedle patch stable at room temperature. *Drug Deliv. Transl. Res.* **2015**, *5*, 360–371. [CrossRef]
26. Chen, M.C.; Lai, K.Y.; Ling, M.H.; Lin, C.W. Enhancing immunogenicity of antigens through sustained intradermal delivery using chitosan microneedles with a patch-dissolvable design. *Acta Biomater.* **2018**, *65*, 66–75. [CrossRef]
27. Caudill, C.; Perry, J.L.; Iliadis, K.; Tessema, A.T.; Lee, B.J.; Mecham, B.S.; Tian, S.; DeSimone, J.M. Transdermal vaccination via 3D-printed microneedles induces potent humoral and cellular immunity. *Proc. Natl. Acad. Sci. USA* **2021**, *118*, e2102595118. [CrossRef]

28. Li, S.; Xia, D.; Prausnitz, M.R. Efficient Drug Delivery into Skin Using a Biphasic Dissolvable Microneedle Patch with Water-Insoluble Backing. *Adv. Funct. Mater.* **2021**, *31*, 2103359. [CrossRef]
29. Yang, Y.; Chu, H.; Zhang, Y.; Xu, L.; Luo, R.; Zheng, H.; Yin, T.; Li, Z. Rapidly separable bubble microneedle patch for effective local anesthesia. *Nano Res.* **2022**, *15*, 8336–8344. [CrossRef]
30. Zhu, L.; Zhuang, J.; Zhao, Z.; Huang, Y.; Sun, J.; Liu, F.; Xue, Q.; Wu, D. Preparation and transdermal effect of a tip-loaded bubble-soluble microneedle. *Sheng Wu Gong Cheng Xue Bao* **2022**, *38*, 2281–2291. [CrossRef]
31. Ning, X.; Chen, S.; Yang, Y.; Hwang, J.; Wiraja, C.; Zhang, C.; Liu, W.; Liu, L.; Xu, C. Photodynamic Bubble-Generating Microneedles for Enhanced Transdermal Cancer Therapy. *ACS Appl. Polym. Mater.* **2021**, *3*, 6502–6512. [CrossRef]
32. Wang, Q.L.; Zhu, D.D.; Liu, X.B.; Chen, B.Z.; Guo, X.D. Microneedles with Controlled Bubble Sizes and Drug Distributions for Efficient Transdermal Drug Delivery. *Sci. Rep.* **2016**, *6*, 38755. [CrossRef] [PubMed]
33. Zhao, Y.; Kang, L.; Gao, S.; Zhou, Y.; Su, L.; Xin, W.; Su, Y.; Wang, J. Expression and purification of functional Clostridium perfringens alpha and epsilon toxins in Escherichia coli. *Protein Expr. Purif.* **2011**, *77*, 207–213. [CrossRef] [PubMed]
34. Yao, W.; Kang, L.; Gao, S.; Zhuang, X.; Zhang, T.; Yang, H.; Ji, B.; Xin, W.; Wang, J. Amino acid residue Y196E substitution and C-terminal peptide synergistically alleviate the toxicity of Clostridium perfringens epsilon toxin. *Toxicon* **2015**, *100*, 46–52. [CrossRef]
35. Gossmann, R.; Langer, K.; Mulac, D. New Perspective in the Formulation and Characterization of Didodecyldimethylammonium Bromide (DMAB) Stabilized Poly(Lactic-co-Glycolic Acid) (PLGA) Nanoparticles. *PLoS ONE* **2015**, *10*, e0127532. [CrossRef]
36. Li, W.; Terry, R.N.; Tang, J.; Feng, M.R.; Schwendeman, S.P.; Prausnitz, M.R. Rapidly separable microneedle patch for the sustained release of a contraceptive. *Nat. Biomed. Eng.* **2019**, *3*, 220–229. [CrossRef]
37. Zhao, B.; Jin, Z.; Yu, Y.; Li, Y.; Wang, J.; Wan, W.; Hu, C.; Li, X.; Li, Y.; Xin, W.; et al. A Thermostable Dissolving Microneedle Vaccine with Recombinant Protein of Botulinum Neurotoxin Serotype A. *Toxins* **2022**, *14*, 881. [CrossRef]
38. Bai, X.; Hu, C.; Jin, Z.; Wan, W.; Li, Y.; Wang, J.; Li, Y.; Gao, S.; Wang, J. Establishment of a thermostable direct hemolysin DAS-ELISA detection method for Vibrio para-haemolyticus. *Chin. J. Zoonoses* **2022**, *38*, 666–672.

toxins

MDPI

*Review*

# Recent Developments in Vaccine Design: From Live Vaccines to Recombinant Toxin Vaccines

**Sonal Gupta and Sabine Pellett ***

Department of Bacteriology, University of Wisconsin-Madison, Madison, WI 53706, USA; sgupta279@wisc.edu
* Correspondence: sabine.pellett@wisc.edu

**Abstract:** Vaccines are one of the most effective strategies to prevent pathogen-induced illness in humans. The earliest vaccines were based on live inoculations with low doses of live or related pathogens, which carried a relatively high risk of developing the disease they were meant to prevent. The introduction of attenuated and killed pathogens as vaccines dramatically reduced these risks; however, attenuated live vaccines still carry a risk of reversion to a pathogenic strain capable of causing disease. This risk is completely eliminated with recombinant protein or subunit vaccines, which are atoxic and non-infectious. However, these vaccines require adjuvants and often significant optimization to induce robust T-cell responses and long-lasting immune memory. Some pathogens produce protein toxins that cause or contribute to disease. To protect against the effects of such toxins, chemically inactivated toxoid vaccines have been found to be effective. Toxoid vaccines are successfully used today at a global scale to protect against tetanus and diphtheria. Recent developments for toxoid vaccines are investigating the possibilities of utilizing recombinant protein toxins mutated to eliminate biologic activity instead of chemically inactivated toxins. Finally, one of the most contemporary approaches toward vaccine design utilizes messenger RNA (mRNA) as a vaccine candidate. This approach was used globally to protect against coronavirus disease during the COVID-19 pandemic that began in 2019, due to its advantages of quick production and scale-up, and effectiveness in eliciting a neutralizing antibody response. Nonetheless, mRNA vaccines require specialized storage and transport conditions, posing challenges for low- and middle-income countries. Among multiple available technologies for vaccine design and formulation, which technology is most appropriate? This review focuses on the considerable developments that have been made in utilizing diverse vaccine technologies with a focus on vaccines targeting bacterial toxins. We describe how advancements in vaccine technology, combined with a deeper understanding of pathogen–host interactions, offer exciting and promising avenues for the development of new and improved vaccines.

**Keywords:** recombinant vaccines; vaccine; toxin; toxoid; botulism; tetanus; COVID-19; mRNA vaccines

**Key Contribution:** This review describes the development of vaccines from early live inoculations to modern recombinant protein and mRNA vaccines, with a particular focus on vaccines to protein toxins.

## 1. Introduction

One of the most effective methods of preventing diseases in humans and animals is the population-wide use of vaccines [1,2]. A vaccine is a biological formulation designed to give rise to a protective and ideally long-lasting immune response against a pathogen without causing disease. Many different approaches have been employed in developing vaccines. Most commonly, vaccines consist of a killed or attenuated form of the entire pathogen, or a surface protein or toxin derivative. All vaccines prime and stimulate the host's immune system to recognize the pathogen as a threat and rapidly eliminate it from the system, hence, protecting the host from severe disease after infection with a pathogen or after intoxication. Vaccines can be prophylactic or therapeutic, where prophylactic vaccines are given prior to encounter of the pathogen, and therapeutic vaccines are administered after encounter of the pathogen [1,2].

The history of vaccines can be traced back to the late 18th century when smallpox was rampant and caused many deaths across Europe. After observations that smallpox survivors were subsequently immune to the disease, human inoculations with a lancet dipped into a ripe smallpox pustule (variolation) became commonplace. While this practice was protective, side effects were serious and frequent, with inoculated persons usually developing mild disease and 2–3% dying from smallpox [3]. In addition, variolated people sometimes caused new outbreaks of smallpox. Towards the end of the 18th century, Edward Jenner, who himself received variolation at the age of 8, noticed that milkmaids who had contracted cowpox, a similar but milder disease, did not get smallpox [4,5]. In an effort to reduce the side effects of variolation, Jenner in 1796 conducted an experiment in which he inoculated a young boy with a fresh cowpox pustule and later exposed him to smallpox. The boy did not get sick, proving that cowpox provided immunity against smallpox. Jenner called this procedure vaccination, based on the Latin word vacca for cow, and vaccinia for cowpox. It took almost 45 years before vaccination eventually replaced variolation in 1840, and after multiple rounds of vaccine alterations to improve safety, the smallpox vaccine eventually developed into one of the greatest medical success stories of humankind, when the disease was eradicated globally in 1977 [3]. Following the initial success of the smallpox vaccine, other vaccines have been developed for a wide range of diseases including measles, mumps, rubella, polio, tetanus, diphtheria, pertussis, hepatitis B, and human papillomavirus (HPV), among others. Emil von Behring and Shibasaburo Kitasato developed the diphtheria antitoxin in the 1890s [4]. Vaccines for pertussis (whooping cough) and tuberculosis were developed in the 1920s [4,5]. Jonas Salk developed the first polio vaccine in the 1940s, which was later replaced by Albert Sabin's oral polio vaccine. Later, in the 1950s, vaccines for measles, mumps, and rubella were developed [4,5]. The first vaccines for hepatitis B were developed in the 1960s. These developments of vaccines has greatly reduced the incidence and severity of human diseases and has saved countless lives.

The smallpox vaccine was developed without an understanding of the virus underlying the disease. Currently, our detailed organismal, molecular, and structural insights into various pathogens enable highly sophisticated and targeted approaches in vaccine developments, dramatically improving the safety of vaccines. Based on the pathogen against which a vaccine is designed (bacteria or virus), there are multiple vaccine technologies to generate a protective immune response (Figure 1). Live attenuated vaccines consist of live pathogens from either a bacterium or a virus that has been weakened through heat or chemical methods. These vaccines contain live pathogens that produce a mild infection but do not result in severe disease, thereby inducing protective cells mediated as well as humoral immunity with strong, effective, and long-lasting protection after only one inoculation [6]. However, one of the major drawbacks of live attenuated vaccines is the risks associated with these kinds of vaccines, including a risk of reversion of the vaccine strain to a virulent strain, leading to disease. In addition, certain populations such as immunocompromised people may be more susceptible to developing disease from an attenuated pathogen. Killed vaccines overcome this issue by not containing any live pathogens. These vaccines, however, still contain the entire pathogen, thus displaying the same immune epitopes capable of initiating a robust immune response. However, killed vaccines usually are less effective than live vaccines in inducing long-lasting protective immunity [6] and require booster shots along with adjuvants to stimulate a strong immune response. Both attenuated and killed vaccines are comparatively inexpensive to make and can be produced on a mass level.

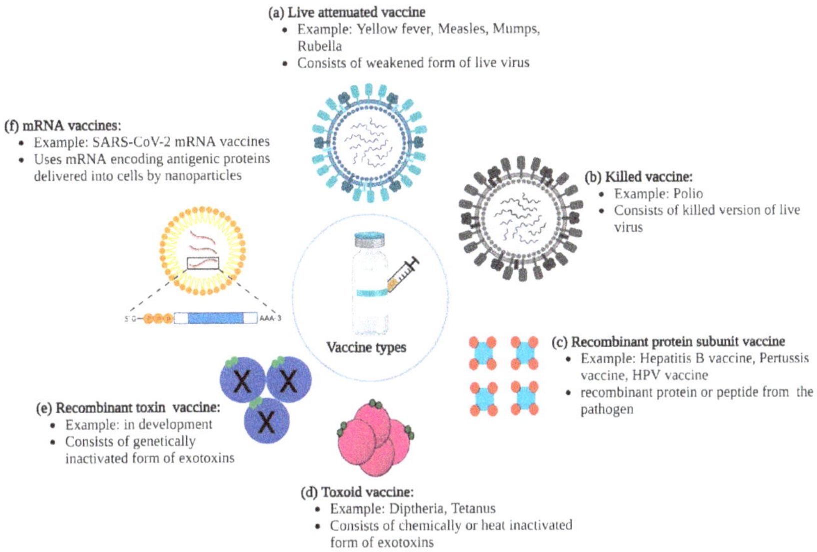


**Figure 1.** Description of different types of vaccine: (**a**) live attenuated vaccine; (**b**) killed vaccine; (**c**) recombinant protein subunit vaccine; (**d**) toxoid vaccine; (**e**) recombinant toxin vaccine; (**f**) mRNA vaccine. Image created with BioRender.com (accessed on 1 August 2023).

With the emergence of genetic engineering technology in the late 20th century, approaches towards vaccine formulation included the development of recombinant protein vaccines [7]. Recombinant subunit vaccines consist of recombinantly produced subunits of a pathogen that stimulates the host's immune system, such as surface proteins or polysaccharides, along with effective adjuvants and, if needed, conjugate carriers [8–10]. An important advantage of this technology is that these vaccines consist of only a part of the pathogen rather than the whole organism, and thus they cannot contain or revert to a live pathogen. The first recombinant subunit vaccine was produced in the mid-1980s against hepatitis B, followed by developments of vaccines against human papillomavirus (HPV), flu, herpes zoster, and coronavirus [9,11,12]. Some antigens of pathogens, including polysaccharides and small peptides, were found to be weak in eliciting a protective immune response due to small size or other structural constraints. Such antigens benefit from formulation as conjugate subunit vaccines, integrating a weak antigen with a strong antigen such as a large protein, as a carrier. Recombinant subunit vaccines are restricted in that they elicit an immune response only against the provided antigen, not the entire pathogen. This can potentially result in weakening protection against pathogens that develop mutations in the protein used as an antigen, and a need for more frequent developments of updated vaccines. While these vaccines generally have improved safety profiles compared to whole pathogen vaccines, the antigens themselves theoretically could retain a potential to cause adverse effects.

A similar strategy used for recombinant subunit vaccines is also applicable to vaccines targeting protein toxins. Some pathogens produce exotoxins that contribute to or cause disease, such as tetanus toxin, botulinum toxin, and diphtheria toxin [13]. The toxin's pathogenic effects can be lessened in severity by immunization with toxoid vaccines, which mount a protective immune response. Toxoid vaccines consist of chemically inactivated toxin mixed with an adjuvant [14]. Multiple boosters are usually required to achieve moderately long-term protection. Newer developments utilize structure–function and pathogenicity knowledge to create non-toxic sub-domains or genetically inactivated toxins as vaccines to increase safety during production and decrease side effects related to residual formalin used in toxoid preparation [15–17].

One of the most recent vaccine strategies that has been in development for the past 3 decades uses deoxyribonucleic acid (DNA) or messenger ribonucleic acid (mRNA) encoding antigens as vaccine candidates [18,19]. This strategy builds on the protein subunit vaccines, but instead of producing recombinant proteins followed by purification and formulation with adjuvants, nucleic acid vaccines use a carrier to deliver DNA or mRNA encoding the proteins to host cells, where the proteins are then produced by the host and released into the circulation to elicit an immune response. The mRNA vaccine strategy received a huge boost in 2020 when US and European governments supported developments of mRNA-based coronavirus disease (COVID) vaccines in an effort to quickly produce enough vaccine doses to immunize large parts of the population in an effort to combat the COVID pandemic.

With so many different vaccine designs, it is tempting to make claims about one technology being better than another. However, many factors need to be considered in choosing the best vaccine approach for each pathogen. Irrespective of vaccine design and formulation, much research and progress remains to be made in the vaccine field, and there's a constant need to develop new vaccines and vaccine approaches. To date, there are no protective vaccines available against fatal diseases such as malaria, shigella, dengue, cholera, and paratyphoid fever. Other common diseases, such as flu and coronavirus, require new developments each year to protect against mutating viruses. This review will discuss various vaccine approaches with a focus on protein vaccines, using prominent examples such as Tetanus, Botulism, and Coronavirus Disease 2019 (COVID-19) to highlight the advantages and disadvantages of various approaches.

## 2. Live Attenuated Vaccines

Since live pathogen inoculation was banned in 1840, live attenuated or inactivated pathogens were the first vaccines developed and widely distributed. These vaccines proved to be one of the most cost-effective and successful methods to prevent human disease, and the approach continues to be used today, including the mumps, measles, and rubella (MMR) vaccine and the chickenpox vaccine. Live attenuated vaccines are known to elicit a broad immune response with efficient and rapid production of protective antibodies produced by B cells as well as cellular CD4+ and CD8+ T lymphocyte (T-cells) responses. This robust immune response usually leads to long-lasting protection and immune memory, potentially life-long, with only one or two administrations. Thus, there often is no requirement for multiple booster administration [20]. The immune response generated by live attenuated vaccines mimics the immune response to live pathogen infection. This includes the production of antibodies by B cells as well as the activation of a cellular immune response involving T cells. Both B-cells and T-cells can differentiate into memory cells, which "remember" the pathogen and allow for a quicker and more effective response in the event of future exposure. Development of B cell memory (long-lived plasma cells or memory B cells) proceeds via two consecutive phases, with both T cells and antigen complexity playing an important role in guiding the development of long-lived plasma cells and memory B cells from naïve B cells [21]. As lifelong protection against disease is the most desirable outcome of vaccination, live attenuated vaccines are an excellent choice for some pathogens. However, there are also several drawbacks to these vaccines. When injected, the attenuated pathogen can theoretically revert to a pathogenic strain at a very low frequency, although this has only been observed for the polio vaccine [22]. Since these vaccines will cause an infection, though much milder than the active pathogen, systemic adverse reactions such as fever, headache, rash, myalgia, etc., are commonly experienced by people after immunization. Side effects may be more severe in immunocompromised or chronically ill [22] individuals. The vaccines also need to be kept cold, thus, making travel and global distribution challenging. The MMR vaccine is one of the most widely used live attenuated vaccines today. Originally created as three distinct vaccines, weakened strains of measles, mumps, and rubella were successfully combined into a single vaccine in 1971 [23]. The MMR vaccine and a quadrivalent vaccine, which includes live attenuated varicella virus, are highly effective as demonstrated by high rates of seroconversion in vaccinated populations (measles: 97.1%, mumps: 96.0%, rubella: 98.8%, and varicella: 93.5%) [24]. In the case of the MMR vaccine, the incidence of serious side effects is about 0.03–0.005% according to the Centers for Disease Control (CDC) and a large study in Finland [25].

## 3. Killed Vaccines

After the huge success of live attenuated vaccines, which eradicated one of the worst infectious diseases of humanity, smallpox, there was an attempt to develop equally cost-efficient and effective vaccines with a greater safety profile, by completely killing the pathogen. Killed vaccines are unable to reproduce inside the host and, as a result, cannot spread, revert to a harmful state, or transmit the disease to others [26]. However, the killed pathogens in killed vaccines proved less effective at inducing a protective immune response and memory, often necessitating formulation with adjuvant systems to increase immunogenicity, and multiple doses to achieve long-lasting protection [26]. Some examples of killed vaccines include the polio vaccine [27] (inactivated poliovirus vaccine, IPV) and the rabies vaccine (inactivated rabies virus vaccine) [27,28]. These vaccines are produced by completely killing the infectious agent through various methods such as heat, chemicals, or radiation, which renders them unable to replicate and cause disease [26]. Inactivated vaccines improved the safety profile of vaccines. For example, the attenuated (oral) Polio vaccine first developed by Sabin in 1961 (OPV) resulted in virus reversion and vaccine-induced paralytic poliomyelitis in 1 in 1.4–3.4 million first vaccine doses administered [29]. To avoid these serious adverse events, the OPV vaccine was replaced with the inactivated polio vaccine developed by Salk in 1955 (IPV) in many countries, and no serious adverse events have been attributed to the IPV vaccine. However, the IPV vaccine also has several disadvantages. It costs five times as much to produce, and it primarily induces a humoral immune response and requires boosts to prime the immune system for memory [30]. More importantly, it induces a much lower mucosal immune response in the intestinal tract, which allows the polio virus to infect the intestinal tract of vaccinated individuals without causing systemic disease but allowing the virus to be shed in their feces, thereby contributing to virus spread during outbreaks.

Finally, manufacturing of killed vaccines involves large-scale in vitro culturing of the causative microorganism. Stringent regulatory requirements make it increasingly challenging to gain acceptance for the development of new vaccines of this type for human use [31]. In addition, one of the challenges of producing inactivated vaccines is to ensure effective inactivation while also preserving protective antigens [26]. In spite of the success of the inactivated polio vaccine, severe adverse events including allergic reactions to vaccine components such as vaccine antigen, preservatives, stabilizer or residual animal protein, and neurologic reactions remain a concern for inactivated vaccines. For the rabies vaccine, severe allergic reactions are observed in 3–6% of patients given the entire three-boost schedule, and 6 cases of neurologic adverse events have been reported including 5 cases of Guillaine–Barre syndrome, according to the CDC [32].

## 4. Toxoid Vaccine

Several human pathogens produce exotoxins, and vaccines against the exotoxins have proven invaluable for public health [4]. Toxoid vaccines have been used since the 1920s and are known to elicit a strong and protective humoral immune response after initial vaccination [33,34], but several booster shots are required to develop longer-term immunity [33,34]. Unlike a vaccine against an entire organism or virus particle, which contains multiple immunogenic epitopes including antigens derived from surface protein epitopes, viral capsids, bacterial cell walls, glyco-epitopes, and polysaccharide epitopes, a toxoid vaccine contains a single protein. However, within the single protein, multiple antigenic epitopes are present, some of which are more immunogenic than others. The process of turning a biologically active toxin into a toxoid usually involves prolonged chemical treatment with formalin to inactivate the toxin. When the toxoid vaccine is injected into the body, it is recognized as foreign and results in a strong humoral immune response, producing antibodies directed against specific antigenic epitopes preserved on the toxoid structure. With repeated boosts, the immune system develops a moderate humoral memory response lasting for several years, allowing for a rapid and robust response with future exposure. However, repeated boosts are required periodically to maintain lifetime protection. Toxoid vaccines can also, to a lesser extent, stimulate a T cell response contributing to protection against the disease [35]. When a toxoid vaccine is administered, the antigen-presenting cells (APCs) such as dendritic cells generally process the toxoid and present peptide antigens on their cell surface using major histocompatibility complex class II (MHC-II) molecules. These MHC-II-peptide complexes act as a signal for T cells, specifically CD4+ T cells (helper T cells) and trigger a cascade of immune responses, resulting in the activation of B cells and cytokine secretion in the host body, ultimately leading to differentiation of B cells to antibody-producing plasma cells and memory B cells [36] (Figure 2). Prime examples of toxoid vaccines are tetanus and diphtheria vaccines, which use chemically inactivated toxoid combined with an adjuvant such as Alum [37] to effectively protect against the disease and will be discussed in more detail below.

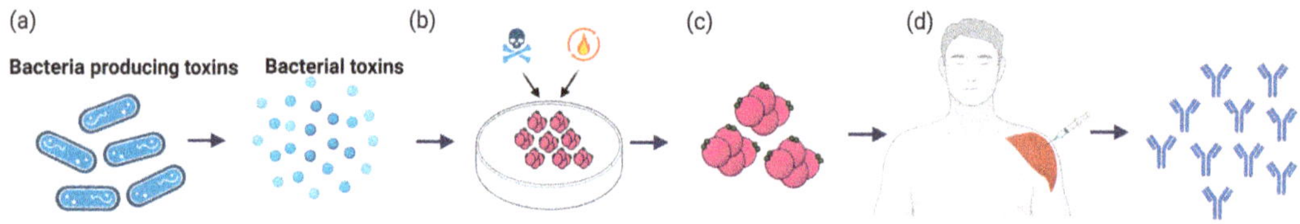


**Figure 2.** Schematic representation to understand the formulation and working of toxoid vaccines: (**a**) The pathogenic bacteria secrete toxins (exotoxins) that contribute to disease, (**b**) Toxins are inactivated or attenuated using heat or formalin resulting in creation of toxoids, (**c**) The toxoids are absorbed with adjuvants to increase immune response, and (**d**) The host body generates antibodies in response to toxoid, providing immunity against the toxin. Image created with BioRender.com (accessed on 1 August 2023).

### 4.1. Tetanus Toxoid Vaccines

Tetanus, a neuromuscular syndrome, is caused by the tetanus neurotoxin produced by the anaerobic, Gram-positive, spore-forming bacterium *Clostridium tetani* after deep wound infection [38–40]. Clostridium tetani produces tetanus toxin (TeNT) as a single-chain 150 kDa toxin that is proteolytically converted to the active di-chain toxin consisting of N-terminal protease light chain (LC) linked by a disulfide bond to C-terminal heavy chain (HC). The HC further consists of a receptor-binding domain ($H_C$) and a LC translocation domain ($H_N$) [41] (Figure 3). TeNT specifically binds to host neurons via polysialoganglioside (GT1b, GD1b) interaction, leading to endocytosis of the entire toxin. The toxin then undergoes retrograde and anterograde transport to the central nervous system, where it enters inhibitory interneurons via transcytosis [42]. In the inhibitory neurons, the LC is released into the cell cytosol via $H_N$-mediated translocation. The LC then cleaves neuronal vesicle-associated membrane protein 2 (VAMP2), which is an essential part of the exocytosis machinery of neurons, and thereby inactivates neurotransmission from inhibitory neurons. This results in over-firing of the activating neurons and associated spastic paralysis [42].

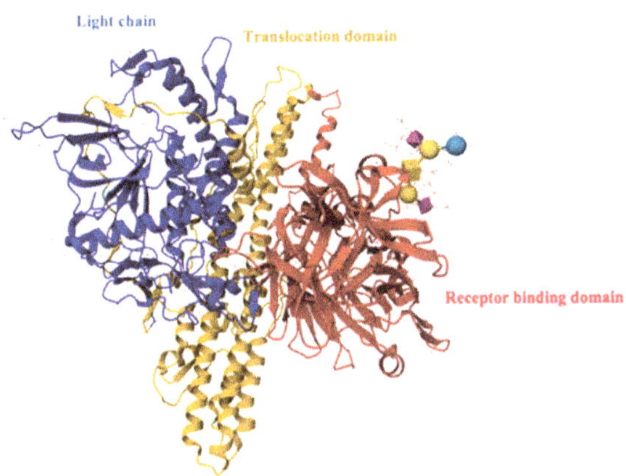


**Figure 3.** Crystal structure of the tetanus neurotoxin (5N0B): crystal structure of the tetanus neurotoxin in complex with GD1a. The image was obtained from the Protein Data Bank (PDB), ID 10.2210/pdb5N0B/pdb [40].

*C. tetani* spores are found in soil, dust, manure, and saliva. Exposure of wounds, deep cuts, or burns to the spores can result in germination and vegetative growth of the bacteria in the anaerobic environment of the deep wound, accompanied by TeNT production, which spreads throughout the body via the circulatory system [38–40]. Tetanus once was a major global disease and today remains a significant threat in many parts of the world, specifically in low-income nations because of lower immunization coverage and unsafe or unhygienic birth practices [36–38]. In 2015, the Global Burden of Disease Study indicated that around 56,000 deaths per year were caused by tetanus worldwide. Treatment for this disease involves administering supportive care, passive immunization with tetanus immunoglobulin, and eradicating bacteria at the site of the wound. Once tetanus develops, the disease causes extreme pain and muscle spasms so strong that they may result in the breakage of bones. The case fatality rate of tetanus remains high (up to 85%), even in areas with abundant medical resources. However, currently, TeNT toxoid vaccines approved by the World Health Organization are included in regular immunization schedules on a global level, dramatically reducing the incidence of tetanus. The TeNT vaccine was first developed in 1924 and is now part of the recommended immunization schedule for children, as it has proven to be effective in preventing tetanus. As of 2018, approximately 85% of infants worldwide (116.3 million infants) had received three doses of the combined diphtheria–tetanus–pertussis vaccine.

The tetanus–toxoid vaccine is based on formalin-inactivated full-length tetanus toxin. Although efficient, these vaccines require complex manufacturing procedures along with pollution and production risks associated with formaldehyde use [43,44]. In addition, large batches (typically 1000 L) of *C. tetani* strain Harvard are fermented for 4–7 days, allowing for the vegetative growth of the bacteria accompanied by active toxin production. The active toxin is then isolated from the bacteria and precipitated by multiple biochemical steps. The yield of semi-purified active TeNT per liter of culture is around 1–10 mg/L, meaning vaccine producers handle extremely large amounts of this potentially lethal and potent toxin, which has an estimated human lethal dose of 0.2 ng/kg intramuscularly [45]. The semi-purified active TeNT is then subjected to a formaldehyde inactivation step, which involves incubation with 0.2–1% formaldehyde for a period of around 6 weeks. This inactivation step results in a completely inactivated toxoid, which can be used for vaccination [46]. While formaldehyde inactivation of protein toxins sounds simple, studies have shown complex chemical modifications resulting in a heterogeneous mixture of adducts and cross-links, which can alter the displayed epitopes and antigen processing [47–49]. Therefore, there is a crucial need for good batch-to-batch consistency, as variation in the antigen structure due to detoxification may result in different efficiency of the final vaccine batch [47]. While formaldehyde treatment generally increases the immunogenicity and stability of the respective protein overall, the extensive chemical alterations occurring to the proteins combined with potentially increased susceptibility to proteolytic degradation [50] potentially result in changes in immunogenic epitopes. It is currently unknown whether such changes may affect the efficacy of the vaccine. Thus, it remains to be elucidated whether formaldehyde treatment is beneficial or detrimental to the immunogenicity of specific protein toxins.

TeNT toxoid vaccines usually are combined with other vaccines, such as the DTaP vaccines (diphtheria, tetanus, and pertussis), Td vaccines (tetanus and diphtheria), and DT vaccines (diphtheria and tetanus). The diphtheria, pertussis and tetanus (DPT) vaccine is a trivalent vaccine consisting of DT toxoid, TeNT toxoid, and either killed whole cells of *Bordetella pertussis* or pertussis antigen [51]. The CDC (Centre for Disease Control) suggests completion of a series of DTaP or DT between ages 6 weeks and 6 years (at 2 months, 4 months, 6 months, between 15 and 18 months, and between 4 and 6 years of age), a dose of Tdap or Td at age 11 to 12 years, and subsequent boosters every 10 years. There is a need for regular boosters of the TeNT vaccine about every 10 years due to the vaccine eliciting only a moderate memory response [52]. As a result, the TeNT vaccines are often unsuccessful in low-income and developing nations, which already suffer from low vaccination coverage, due to an unwillingness to receive multiple doses and boosters [53,54]. Thus, there is interest in the development of a new vaccine against tetanus neurotoxin, ideally with lifelong protection from one vaccination schedule.

### 4.2. Diphtheria Toxoid Vaccines

Diphtheria, a severe respiratory infection, is primarily caused by toxigenic strains of the Gram-positive, invasive bacterium *Corynebacterium diphtheria* [55]. While *C. diphtheriae* is an invasive bacteria that enters respiratory tissue host cells via adhesins, haemagglutinins, and surface-exposed non-fimbrial proteins, which contributes to pathogenicity, a major virulence factor causing host cell death is the secreted diphtheria exotoxin (DT). DT is an elongation factor 2 (EF2) adenosine diphosphate (ADP) ribosylating toxin [56]. DT is a 58 kDa polypeptide that is secreted across the cytoplasmic membrane of *C. diphtheria* without cell lysis [55]. Like CNTs, DT is a disulfide-linked dichain protein consisting of the catalytic (C) subdomain and a carboxyl-terminal subdomain containing the translocation (T) and receptor-binding (R) domains. The R domain of DT binds heparin-binding epidermal growth factor (HB-EGF) receptors on human epithelial cells, leading to clathrin-mediated endocytosis of the entire toxin-receptor complex. The acidic conditions in the endosome trigger a conformational change leading to membrane insertion of the T domain and translocation of the C domain into the cell cytosol, where it catalyzes the transfer of ADP from nicotinamide adenine dinucleotide (NAD) to EF-2 [55]. Since EF-2 is an essential factor for cell protein biosynthesis, ADP ribosylation by DT prevents proper EF-2 function, thereby shutting down host cell protein synthesis ultimately leading to cell death [55].

It is estimated that before the 1980s, around one million diphtheria cases occurred annually on a global level. The diphtheria toxoid vaccine, which is prepared similarly to tetanus toxoid by inactivating the toxin with formalin, was first developed in 1920 and was successful in preventing serious disease. Vaccine use decreased global diphtheria cases to ~100,000 by 1980, when more widespread use of the vaccine resulted in a strong and steady decline in case numbers [55]. In 2015, 4500 cases were officially reported, a notable decrease from the nearly 100,000 cases recorded in 1980 according to WHO reports [57]. Presently, diphtheria is most prevalent in sub-Saharan Africa, India, and Indonesia. Although rare in developed countries due to widespread vaccination, there is a risk of resurgence if vaccination rates decline. Between 1980 and 2004, the United States reported only 57 cases.

The immune response stimulated by diphtheria toxoid is similar to that stimulated by tetanus toxoid. Like tetanus toxoid, diphtheria toxoid vaccines are relatively cost-effective to produce, but have several downsides including the need for the production of a potent and dangerous toxin at a large scale, differences in the purity of the protein toxin due to varying purification processes, the long chemical inactivation procedure leading to batch to batch differences, and the need to use formalin.

### 4.3. *Botulinum Neurotoxin Toxoid Vaccines*

BoNTs (botulinum neurotoxins) are the most potent protein toxins known to humans to date [58]. There are seven immunologically distinct BoNT serotypes (A–G), with natural variants within each serotype termed subtypes [59]. BoNTs share overall structural similarity with TeNT. They also are 150 kDa protein toxins that are proteolytically processed to form a 100 kDa heavy chain (HC) and a 50 kDa light chain (LC) linked via a disulfide bond (Figure 4). Unlike TeNT, which is produced as a single 150 kDa polypeptide, BoNTs are produced by their clostridial hosts in complex with several non-toxic proteins (NTNH (nontoxic non-hemagglutinin (NTNH)) and HA (hemagglutinin) or orfx proteins). The LCs are zinc proteases that specifically cleave plasma membrane or vesicle-associated SNARE (soluble N-ethylmaleimide-sensitive factor activating protein receptor) proteins [60]. The HC can further be functionally and structurally divided into an $H_N$ domain (translocation domain) aiding the LC in entering the cell cytosol and an $H_C$ domain (Receptor-binding domain) that specifically associates with ganglioside and protein receptors on neuronal cells [61,62].

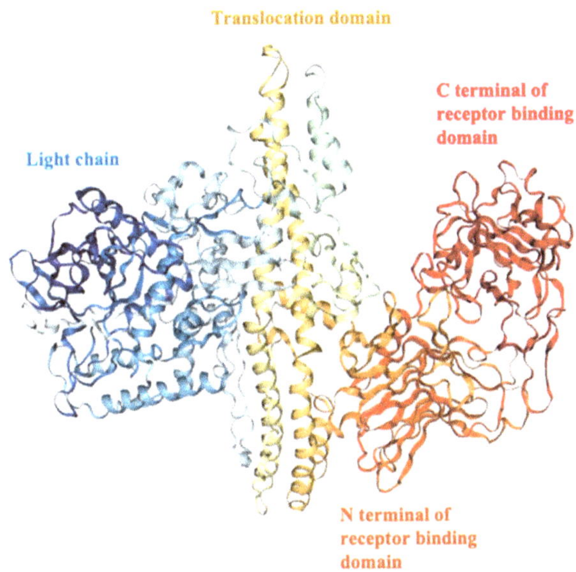


**Figure 4.** Crystal structure of botulinum neurotoxin (3BTA): crystal structure of botulinum neurotoxin serotype A. The image was obtained from the Protein Data Bank (PDB), ID 10.2210/pdb3BTA/pdb [63].

BoNTs primarily target motor neurons, and unlike TeNT, the BoNT LCs are released directly into the cytosol of the motor neurons, where they cleave SNAREs at the neuromuscular junction, resulting in inhibition of neurotransmitter release by the motorneurons. This leads to long-lasting flaccid paralysis. Once inside a neuron, BoNT LC can stay active for a prolonged time of days to several months depending on the BoNT serotype [64–66]. A vaccine to BoNTs, thus, has to be highly efficient in removing the toxins from the circulation before neuronal cell entry. The major five kinds of botulism are infant botulism, wound botulism, foodborne botulism, iatrogenic botulism, and very rarely, adult intestinal toxemia [67,68]. Wound botulism can occur when spores of botulinum toxin-producing clostridia enter a deep wound and produce the toxin during growth within the wound, similar to tetanus. Foodborne botulism occurs after consuming botulinum neurotoxin already present in improperly canned, preserved, or fermented homemade foods, as well as occasionally from store-bought foods. Infant botulism typically occurs when infants consume contaminated substances, such as honey or soil, consisting of *Clostridium botulinum* spores. The bacteria further grow and produces toxin in the baby's intestines. Since an infant's digestive system is not fully developed, they are more susceptible to the bacteria's growth and toxin production. Iatrogenic botulism refers to botulism that is caused by medical or healthcare interventions. It is a rare form of botulism that can occur when botulinum toxin is administered for therapeutic or cosmetic purposes but is not properly administered, monitored, or controlled.

No cure exists for botulism, but protective vaccination with botulinum neurotoxin toxoid, prepared by formalin inactivation similar to TeNT and DT toxoids, has been shown to be effective in preventing botulism [69,70]. While vaccinating the general population against BoNTs is not desirable due to the significant clinical benefits provided by pharmaceutical BoNTs, certain populations are more vulnerable to the disease and would benefit from vaccination. Since 1965, the CDC has made an investigational pentavalent BoNT toxoid (PBT) vaccine available to workers at risk of exposure under an investigational new drug (IND) application (BB-IND 161), including laboratory workers and some military personnel [16]. The primary vaccination series included injections at 0, 2, and 12 weeks, followed by annual boosters. Moderate and severe local reactions were reported in 10% and 3% of vaccine recipients after the boosters, respectively, while <0.3% reported systemic reactions (inflammation, itching along with nausea, vomiting, or diarrhea), among the approximately 4200 individuals who received the vaccine [71]. Prior to the mid-1990s, around 50% of vaccine recipients had anti-BoNT antibody levels of 0.2 International Units (IU)/mL or higher at their 2-year post-booster assessment. However, by 2004, less than 15% of vaccine recipients had the same levels of antibodies and there was a decrease in the production of antibodies against toxin serotypes C, D, and E, indicating waning vaccine efficacy. Also, potency studies on vaccine lots manufactured in 1976 and distributed since 1982 showed that they no longer effectively protected guinea pigs from serotypes D and E [62]. Eventually, due to declining potency and efficacy and the high prevalence of local side effects, the pentavalent (ABCDE) botulinum toxoid vaccine was discontinued in 2011 [71–73]. No human vaccine for BoNTs has been approved since. However, there have been multiple efforts for the development of new vaccines, including recombinant protein vaccines against parts of the BoNT toxins or mutated holotoxins with diminished biologic activity [74–78]. Today, there remains a significant requirement for the development of an effective vaccine for BoNTs.

While in the US the main focus for new BoNT vaccine development is using novel approaches of genetically de-toxified or biologically inactive protein subunit vaccines, scientists in Japan developed another tetravalent chemically detoxified vaccine within the last two decades [79]. This vaccine encompasses formalin-inactivated toxoids of BoNT serotype /A, /B, /E, and /F, utilizing the progenitor M toxin complexes (BoNT and NTNH) [70]. A small human clinical study ($n = 43$) showed neutralizing antibodies against all four BoNTs, with antibody titers increasing with each boost and then slowly waning over time [79–81]. While it is unclear why the study used the M-complex including NTNH for vaccine development, this small clinical study looks promising, with 86% of participants reaching protective neutralizing antibody levels above 0.25 IU/mL [70]. In another study in Japan, BoNT/A toxoid combined with a mutated cholera toxin produced BoNT-specific IgG in plasma and IgA in external secretions (saliva, fecal extracts, and nasal washes) in mice [80]. Mice receiving this nasal vaccine were completely protected against challenge with 4000 mouse lethal dose 50 (LD50) Units of BoNT/A by challenge through the intraperitoneal route and against 2 LD50 Units of oral delivered progenitor BoNT/A complex [80]. Taken together the studies on BoNT toxoid vaccines show that chemical detoxification of BoNTs can produce an efficient vaccine that elicits a strong neutralizing humoral immune response to the toxins. However, data on immune memory are limited and yearly boosters were required for most vaccine recipients of the investigational pentavalent BoNT toxoid vaccine to maintain neutralizing antibody levels. Another consideration for BoNT toxoid vaccines is the significant challenges associated with large-scale active toxin production. In contrast to tetanus toxin, against which most people are vaccinated, there currently is no vaccine against BoNTs, and thus the large-scale production of active BoNTs presents a significant biosafety concern for involved laboratory staff. In addition, in the USA, BoNTs are Tier 1 Select Agents, and thus production also involves significant regulatory hurdles associated with high costs. Therefore, newer developments focusing on alternative approaches that do not require large-scale production of the toxins would be of value.

## 5. Recombinant Subunit and Toxin Vaccines

In the last decades, our understanding of molecular mechanisms underlying pathogenesis, as well as genetic methods and protein engineering technology, have advanced dramatically. This enabled the design of recombinant subunit vaccines consisting of specific surface antigens of the pathogens (Figure 5). This approach has the advantage that specific immune-dominant or biologically essential antigens can be used to elicit a strong immune response for effective neutralization and immune clearance of the pathogen. In addition, these vaccines do not involve whole organisms and, thus, completely overcome safety concerns due to potential strain reversion or ineffective pathogen killing [82,83]. On the other hand, subunit vaccines target only the specific protein or subunit provided and lack the myriad of additional epitopes that can be found on whole organisms. They also lack the capacity to invade host cells as the attenuated or killed pathogen vaccines often do, thereby restricting the T-cell immune response. In general, proteins alone do not elicit strong enough immune responses for vaccination; thus, they are formulated with adjuvants to stimulate the immune system. Alum (aluminum salts) is an economical and safe adjuvant approved by the United States Food and Drug Administration (US FDA) for veterinary and human use and is commonly used for protein subunit vaccines. Alum works by forming a short-term depot at the site of injection and slowly releasing antigens to the body's immune response system [84,85]. Additional adjuvants with specific immunomodulatory functions are actively investigated and hold great potential to improve protein subunit and recombinant toxin vaccines by modulating the induced immune response [86–88].

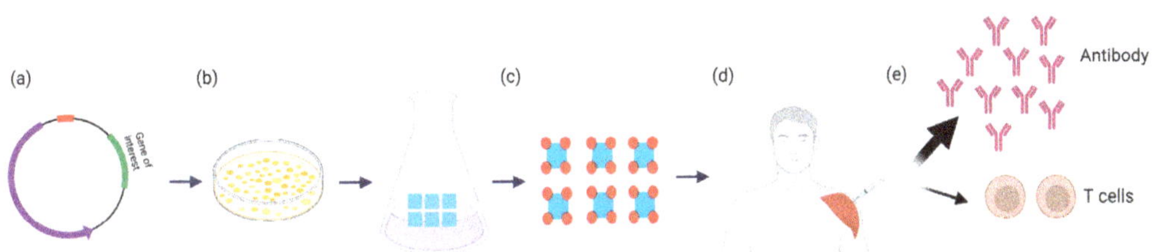


**Figure 5.** Development of recombinant subunit vaccines: (**a**) Plasmid encoding immunogenic genes, (**b**) Plasmid is transformed into expression host cells (bacteria, yeast or animal cells), and antigen is expressed and purified. (**c**) The purified antigenic protein is formulated with adjuvant to boost immune response. (**d**) The vaccine candidate is injected into the host body. (**e**) The injected antigen triggers a host immune response, resulting in generation of antibodies specific to the antigen. Image created with BioRender.com (accessed on 1 August 2023).

Recombinant subunit vaccines are known to elicit strong humoral and CD4+ T cell responses [8,9], along with poor induction of cellular immune responses, particularly CD8+ T cells [8,9]. Therefore, multiple boosters and adjuvants are required to induce immune memory for long duration [89–91]. One well-known example of a recombinant protein vaccine currently used in humans is the vaccine against the hepatitis B virus (HBV). Hepatitis B is a global liver disease, which can result in serious life-long chronic illness. The first HBV vaccine was produced in 1981 and was a killed vaccine derived from the plasma of asymptomatic carriers [92]. Despite a very good safety profile, the high cost of this vaccine combined with the theoretical risk of contamination with other blood-borne viruses such as human immunodeficiency virus (HIV), resulted in the gradual replacement of this vaccine with a recombinant subunit vaccine starting in 1986 [92]. To produce the recombinant vaccine, the hepatitis B surface antigen (HBsAg) is expressed in yeast cells in the absence of any other viral DNA. The resulting recombinant B surface antigen is capable of assembly into virus-like particles (VLPs) of similar size as the virus, which are equally immunogenic and effective as the plasma vaccine [93,94]. Another example of a recombinant vaccine is the vaccine against human papillomaviruses (HPVs), which cause various mucocutaneous diseases, including cervical, vulvar, and vaginal cancers, as well as genital warts [9].

The same vaccine strategy that is applied for recombinant subunit vaccines is also proving valuable in the design of a new type of toxin vaccine, where the toxin is genetically inactivated by the introduction of specific amino acid mutations that render the toxin a biological null. The only currently used example of a recombinant genetically inactivated exotoxin is CRM197. CRM-197 is a genetically detoxified form of the diphtheria toxin (63,000 Da) that is widely used as a carrier protein of conjugated vaccines. To create CRM197, a single mutation at position 52, substituting glutamic acid for glycine, causes the ADP-ribosyltransferase activity of the native toxin to be reduced by about 1 million fold [95,96]. As a result, CRM197 is no longer capable of causing the severe illness associated with diphtheria [95]. However, it retains the ability to stimulate an immune response when used as a component in vaccines and has found great utility as a carrier protein for polysaccharides or small peptides that alone would not be able to induce an immune response, due to its availability of multiple conjugation sites. Examples are the conjugate vaccines to protect against Haemophilus *influenzae* type b or *Streptococcus pneumoniae*. CRM197 has also been conjugated to epidermal growth factor (EGF) to create the EGF-CRM197 vaccine as a potential anti-tumor vaccine for cancers that secrete high levels of EGF. These results are important for future vaccine research and understanding the mechanism of CRM197's anti-tumor effect [97]. Why has CRM197 not replaced the current diphtheria toxoid vaccine? In addition to production cost, caution is warranted when using this recombinant protein at a high dosage due to studies indicating remaining residual toxicity and weak catalytic activity in cultured cells, although >1 million fold lower than the wild-type diphtheria toxin [98]. The human LD50 for DT (diphtheria toxin) has been estimated at 100 ng/kg; thus, for CRM197, one could extrapolate an LD50 would be 100 mg/kg. The dose required to induce immunity is likely >10,000–100,000-fold lower than this estimated LD50. Further research is required to determine the safe dose effectiveness of CRM197 as a protective vaccine compared to diphtheria toxoid. The following sections will discuss similar developments with tetanus and botulinum neurotoxin vaccines.

### *5.1. Recombinant Tetanus Toxin Vaccines*

The current chemically inactivated tetanus toxoid vaccine has been successfully used since 1924 to prevent tetanus, reducing cases by 95%, and continues to be widely used. However, there are several disadvantages associated with this commonly used vaccine. The primary disadvantage is that it requires large-scale production and purification of the highly potent tetanus toxin in its active form before it is inactivated. Even though laboratory workers can be immunized with the current vaccine, reducing the biosafety risk, the large amounts of toxin handled nevertheless present a safety and security risk, and regulatory approvals are required. Only a few manufacturers currently are willing and approved to produce TeNT at the large scale needed for vaccine production, and vaccine shortages occur periodically. Other disadvantages of the current toxoid vaccine include a relatively high incidence of local reactions in 25–85% of patients and severe allergic reactions in ~1 in 100,000 vaccine recipients [35,99,100]. In addition, an initial series of three shots must be followed by periodic boosters about every ten years to maintain immunity [99]. While some improvements to the current tetanus toxoid vaccine may be possible by adjustments in the formaldehyde inactivation process, removal of trace amounts of formaldehyde in the final vaccine, as well as adjuvant formulation, exciting new developments are investigating genetically inactivated recombinant toxin vaccines as an alternative. Research studies have explored atoxic toxin subunit domains, such as the recombinant non-tagged Hc domain of the tetanus toxin expressed in *Escherichia coli* as a recombinant TeNT subunit vaccine [101]. Another recent study developed a genetically detoxified version of tetanus toxin called 8MTT, which retained 99.4% similarity to the original toxin but had eight amino acid mutations to disable several of the intoxication steps of TeNT, including neuronal cell association, translocation of the LC into the cell cytosol, and enzymatic activity of the LC [15]. The modified toxin was purified and after injection into outbred mice at 0.6 mg 8MTT/mice, no symptoms or pathology were observed, suggesting that *E.coli* produced 8MTT is at least 50 million-fold less toxic than native tetanus toxin, which had an LD50 of ~6 pg/mouse in the same study [15]. Vaccination of mice with 8MTT resulted in a strong and robust immune response, which was similar to the immune response in mice vaccinated with an equal dose of chemically inactivated tetanus toxoid after one boost but showed a more pronounced immune response and enhanced survival to toxin challenge after two boosts [15]. In addition, a recent study suggests that 8MTT could also serve as a conjugate vaccine carrier for a small peptide vaccine (tick peptide P0) [102]. Since this peptide is too small to mount an immune response by itself, conjugation to a carrier protein is required. Comparing the immune response in mice immunized with the 8MTT-P0 conjugate versus the chemical TeNT toxoid-P0 conjugate demonstrated an at least equal and strong immune response, validating the 8MTT as a conjugate vaccine carrier. Thus, 8MTT offers an alternative to tetanus toxoid as a primary vaccine and as a conjugate vaccine carrier that is safer to produce, is chemically well defined, and does not require extensive formaldehyde inactivation. Additional studies are needed to further confirm the safety and effectiveness of this approach. Of particular interest would be a combined CRM197—8MTT vaccine.

### 5.2. Recombinant Botulinum Neurotoxin Vaccines

Similar to the development of recombinant TeNT vaccines, this approach is also investigated for the development of recombinant-engineered BoNT vaccines [69]. However, in the case of BoNTs, vaccine development is complicated by the immunogenic diversity of the BoNT serotypes, in effect requiring the development of a multivalent vaccine protecting against all BoNT sero- and subtypes that can cause human botulism. The previously available investigational pentavalent BoNT toxoid vaccine had many downsides associated with it, including the need to produce large amounts of active toxins, reactions to residual formaldehyde in the formulation, varying efficacy for the five included serotypes, the need for yearly boosters, and overall waning efficacy. This eventually led to its discontinuation in 2011 [71], and no approved vaccine has been available since. In addition to serving as a vaccine for at-risk populations, this vaccine also enabled the development and production of Human Botulism Immune Globulin Intravenous (BIG-IV, BabyBIG®), which is derived from donated plasma of vaccinated individuals and enables the safe treatment of infants suffering from infant botulism, to both milden disease severity and reduce disease duration [103]. Discontinuation of the vaccine thereby also resulted in a stop in the supply of plasma for further production of this important drug. Thus, efforts to produce new anti-BoNT vaccines commenced rapidly, and in an effort to avoid the downsides previously encountered with the toxoid vaccine, the new efforts focused on recombinant vaccine developments. Recombinant engineered HC, $LCH_N$, and full-length BoNT expressed in the yeast *Picchia* or *Escherichia coli* have been studied as vaccine candidates in animal studies. BoNT/A1-$H_C$ expressed in *E. coli* had been shown to elicit an antibody immune response against $H_C$, resulting in partial protection from challenge by 1200 LD50 units of BoNT/A [104]. Considering that BoNTs require a functional $H_C$ domain to associate with neuronal cells for cell entry, this approach targeted the most obvious domain of BoNTs that would result in toxin neutralization. Further studies examined additional HC domain constructs produced in yeast or *E. coli* and found consistent protection against botulism in vaccinated mice [105–107].

The $H_C$ of all seven BoNT serotypes (hepta- $H_C$) has been successfully expressed in *E. coli*. When outbred mice were vaccinated with hepta-$H_C$, they developed antibodies against all seven BoNT $H_C$s and were protected from challenge by 10,000 LD50 Units of each BoNT serotype [108]. In a small human clinical study, a recombinant $H_C$ A and B vaccine produced in the yeast *Picchia pastoris* was assessed in a small cohort of human volunteers who had previously been immunized with the pentavalent toxoid vaccine [109]. The study resulted in a satisfactory 4-fold increase in neutralizing antibodies against BoNT/A and BoNT/B in 82 and 87% of study participants, and several participants donated their plasma for continued production of BIG-IV [109].

The $H_C$ vaccines are easy to produce and provide protection against intoxication. However, studies have shown that the LC and $H_N$ domains of BoNTs are more immunogenic than the HC domains [66,110], suggesting that optimal vaccines should include these domains. Other studies have demonstrated that even if the $H_C$ is modified to lack ganglioside binding, it can still stimulate protective immunity in outbred mice [105–107,111], as well as that catalytic activity of the BoNT/LC can be diminished by introducing two to three single amino acid mutations that prevent coordination of the zinc in the LC [74,112,113]. This led to the idea of genetically inactivated holotoxin vaccines that are not able to produce toxicity but maintain their overall structure and immunogenicity. Multiple such BoNT holotoxin constructs with reduced toxicity have been designed and produced in various expression hosts including *Clostridium* [114], *E. coli* [115,116], *Picchia pastoris* [117], and insect cells [118]. All these constructs utilized two to three single amino acid mutations within the LC domain of BoNT to render the toxin catalytically inactive by preventing the coordination of the zinc within the endopeptidase. The toxicity of these constructs was dramatically reduced compared to wild-type BoNTs, ranging from detected residual toxicity of 56 μg/kg to 5 mg/kg in mice, compared to ~150 pg/kg for native toxins [118]. However, as these constructs were produced in heterologous hosts and usually with protein tags to facilitate purification, at least some of the reduction in toxicity may be due to a lack of or sub-optimal post-translational proteolytic processing of the toxin or to protein tags. Since proteolytic processing theoretically can still occur in vitro, including after injection into a host due to host enzymes or by production in a different expression host, further research of these mutant proteins produced in a native expression host is required to directly determine the effect of the individual amino acid substitutions on BoNT toxicity potential.

In a recent study, structure-function knowledge of BoNTs was employed to create holotoxin constructs with distinct amino acid mutations designed to eliminate several biologic functions of BoNT including cell association and LC endopeptidase activity [77,106,119]. The resulting 4M-BoNT/A (mutations) produced in *E. coli* resulted in no detectable toxicity at doses 1 million fold greater than the lethal dose of BoNT/A [120]. However, it is still unclear whether this reduction in toxicity is due to the four introduced single amino acid changes or due to factors related to *E. coli* production and purification.

Several studies evaluated BoNT holotoxin constructs as genetically inactivated recombinant vaccines, comparing vaccine efficacy to toxin subunit vaccines including the $H_C$ domains or LC-$H_N$ domains [117]. In all studies, the full-length genetically inactivated BoNT vaccines resulted in the greatest and quickest production of protective antibodies and the most protection against the BoNT challenge, including multiple subtypes of each serotype. Interestingly, the LC-$H_N$ vaccines resulted in almost equal protection against the BoNT challenge and only slightly reduced overall IgG production [121]. In addition, analyses of the antibodies formed in mice vaccinated with full-length BoNT vaccines showed that the majority of antibodies recognized the LC and $H_N$ domains [122]. This is also in agreement with studies analyzing human sera from botulism patients, which indicated that the LC and $H_N$ domains of BoNTs appeared the most immunogenic [123–125]. However, since antibodies against the $H_C$ domain would be expected to mask receptor binding and thereby neutralize toxin cell entry, the inclusion of the $H_C$ domain into vaccine design might still play an important role. In addition, immune memory has not been analyzed with these vaccines yet and remains an outstanding challenge [74,107,120,126–130]. In terms of production, the $H_C$ and LC-$H_N$ domains are easily produced in *E. coli* and production can be scaled to high levels [107]. Production of stable and soluble holotoxin, on the other hand, has proven challenging, and further work is required to optimize such production. It should also be noted that the LC-$H_N$ constructs have the potential to enter neurons at very high concentrations; thus, the catalytic activity of a LC-$H_N$ vaccine should be eliminated.

Overall, the approach of a recombinant genetically inactivated protein toxin vaccine is promising. Further work is required to both, strengthen our understanding of the molecular mechanisms and structure-function relationships of BoNTs, and to determine mutations resulting in complete elimination of the biologic activities of BoNTs while not altering the overall structure for optimal antigenic epitope display. In addition, direct comparisons of recombinant toxin vaccines with corresponding chemical toxoid vaccines in terms of safety, efficacy, and ease of production will determine the practical utility of this approach. The idea of preserving overall protein structure by avoiding chemical treatments is targeted at creating a vaccine with greater efficacy. However, even if vaccine efficacy is equal to that of the toxoid vaccine, as has been observed for the 8MTT vaccine [15], the advantages of producing an atoxic protein versus a highly potent and regulated neurotoxin are significant from a safety and financial perspective.

### *5.3. Recombinant Subunit Vaccines to SARS-CoV-2*

The global COVID-19 pandemic that started in 2019 required rapid development of an effective vaccine. The coronavirus strain responsible for COVID-19 is severe acute respiratory syndrome coronavirus 2 (SARS-CoV-2) [131,132]. SARS-CoV-2 belongs to the Betacoronavirus genus of the *Coronaviridae* family and shares close genomic similarities with SARS-CoV, an earlier endemic virus that first emerged in 2002–2003. SARS-CoV-2 and SARS-CoV are both positive-sense single-stranded RNA viruses with a genome size of ~30 kilobases that encode several structural and non-structural proteins. Scientists around the globe worked quickly to determine the structure, RNA sequence, and pathogenesis mechanisms of SARS-CoV-2 [133–137]. While during the pandemic, vaccine development focused on the production of mRNA vaccines encoding the binding domain of the S protein, recombinant protein vaccines have also been a popular subject of research [138]. The structural proteins of SARS-CoV-2 contain the spike (S) glycoprotein, envelope (E) protein, membrane (M) protein, and nucleocapsid (N) protein [139–141]. Since the spike proteins from SARS-CoV-2 (S) are crucial in virus attachment and entry into host cells via interaction with the angiotensin-converting enzyme 2 (ACE2) receptor, S proteins emerged as a key target for vaccine development to induce neutralizing antibodies [142]. Subunit vaccines exclusively focus on targeting a specific protein or subunit, disregarding the multitude of other epitopes present in complete organisms. For example, analyses of COVID-19 patient sera showed antibodies to over 800 distinct immunogenic epitopes on the SARS-CoV-2 virus, with only a subset of these present on the spike protein that was used in the widely distributed mRNA or viral vector vaccines [143,144]. Since the spike protein is essential for cellular entry by the virus and also plentifully adorns the surface of the SARS-CoV-2 virus, the specific targeting of the multiple antigenic epitopes present in the spike protein-binding domain provided a strong neutralizing antibody response protecting from severe disease [145]. The spike protein, however, is also a frequently mutated protein of this virus, thus requiring constant updating of the vaccine.

A vaccine candidate based on the SARS-CoV-2 virus spike protein named "NVX-CoV2373" was developed and produced by biotechnology company, Novavax, based in the United States. Novavax had reported positive results from Phase 3 clinical trials conducted in the United Kingdom [146]. This protein subunit vaccine is shown to be capable of providing immunity against SARS-CoV-2 infection [147–149]. However, since the S protein retains its biologic property of being able to bind to the ACE2 receptor, certain sub-populations may be susceptible to deleterious side effects, as has also been observed with the corresponding mRNA vaccine [147,150]. An interesting approach to decrease such a risk of side effects would be introducing specific amino acid mutations that prevent

ACE2 binding but maintain overall structure, similar to the approach used in the design of genetically inactivated protein toxin vaccines.

## 6. mRNA Vaccines

The concept of mRNA vaccines was first introduced in the early 1990s, but technological hurdles and delivery challenges delayed significant progress until recent years. Prior to the COVID-19 pandemic, several mRNA vaccines were already in clinical trials to treat various infectious diseases, including HIV-1, rabies, Zika, influenza, and multiple hematologic and solid organ malignancies [151–153]. During the COVID-19 pandemic, mRNA vaccine technology rose to prominence, and the most widely used early vaccines against SARS-CoV-2 were based on mRNA vaccine technology.

The mRNA vaccines are based on mRNA encoding a bacterial or viral protein that is immunogenic enough to elicit an efficient humoral or cellular immune response in the host body upon administration. The mRNA is packaged within nanoparticles for cellular delivery and is modified such that it can evade the host cell's immune defenses. Once inside the cells, mRNA gets translated by host cell ribosomes to create a protein antigen. In most cases, the protein antigen contains a secretion peptide sequence for efficient release from the host cell into the patient's circulatory system, where the host's immune response mounts an effect leading to an immune response [154] (Figure 6). The mRNA vaccines directly deliver mRNA to the cytoplasm, bypassing the nucleus and avoiding incorporation into the genome. Its presence in cells is short-lived, and it is quickly broken down and eliminated through cellular processes. The mRNA vaccines can be designed, formulated, and mass-produced very rapidly, unlike traditional vaccines that take months to develop [152–154]. The ability to rapidly produce large amounts of vaccine was of great importance during the COVID-19 pandemic. A disadvantage of mRNA vaccines, however, is that they need to be stored at very cold temperatures (−80 °C), requiring logistics and costs associated with a cold storage chain for global distribution. Another disadvantage of these vaccines was the relatively new technology of nanoparticle delivery and the relative lack of safety data. However, after the COVID-19 pandemic, millions of mRNA vaccines have been administered to patients demonstrating a very low incidence of severe side effects and no deaths due to the vaccine.

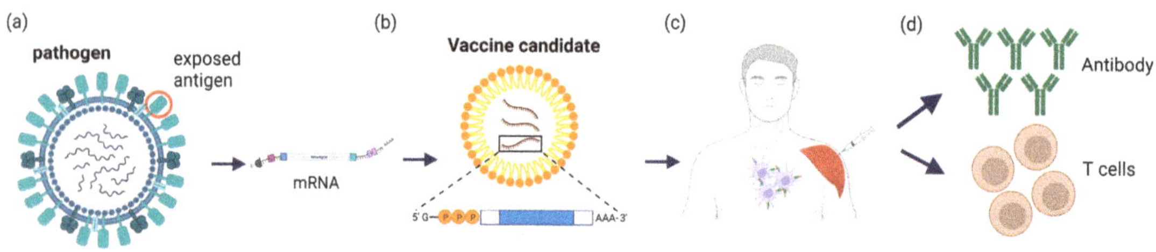


**Figure 6.** An overview of development strategy for mRNA vaccines. (**a**) The gentic sequence from the pathogen antigen codes for synthetic mRNA sequence. (**b**) Synthetic mRNA is packaged into lipid nanoparticle. (**c**) The vaccine is injected into the host. The mRNA is delivered to immune cells, resulting in production of antigen. (**d**) The vaccine stimulates host's immune response resulting in production of antibodies and T cells specific to the antigen. Image created with BioRender.com (accessed on 1 August 2023).

### 6.1. SARS-CoV-2 mRNA Vaccines

Due to the swift worldwide transmission of SARS-CoV-2 in the initial year of the COVID-19 outbreak, the mRNA vaccine approach was adopted as a primary method for expeditious and efficient vaccine production. The vaccines were based on synthetic mRNA strands encoding the SARS-CoV-2 spike glycoprotein, packaged in lipid nanoparticles.

Of the multiple recombinant vaccines approved in the United States against COVID-19, including Pfizer-BioNTech, Moderna, Novavax and Johnson & Johnson's Janssen, the two most frequently used and earliest available were mRNA vaccines. BNT 162b2 (Pfizer) and mRNA-1273 (Moderna) were approved for emergency use by the Food and Drug Administration (FDA) in less than a year [155,156]. Both vaccines were delivered intramuscularly with at least two doses required for protection, and a third dose booster for longer-lasting immunity [155,156]. These vaccines were the first widely used human mRNA vaccines after they were authorized for emergency use in multiple countries around the world. While there was significant debate on the idea of widely distributing these vaccines without the usual safety studies required in vaccine developments, we have now gained great insight into the safety and efficacy of mRNA vaccines after 2 years of widespread applications and over 5.5 billion people, or 72% of the entire world population, having been vaccinated against COVID-19. The side effects associated with these vaccines included pain, tenderness, swelling, or redness at the injection site along with fatigue, headache, muscle pain, chills, joint pain, fever, nausea, malaise, and lymphadenopathy [155,156]. According to clinical trial data and real-world studies, the reported incidence of lymphadenopathy following COVID-19 vaccination varies between 0.3% to 16% of vaccinated individuals. These numbers can fluctuate depending on factors such as the vaccine used and the specific population studied. [157]. In rare cases, a severe allergic reaction occurred within one hour of receiving either vaccine [158,159]. Another serious side effect that has been linked to mRNA-based COVID vaccines is an increased risk of temporary and treatable myocarditis or pericarditis in up to 154 per one million vaccine recipients, particularly in young male populations [160]. However, the risk of developing myocarditis or pericarditis due to COVID-19 infection is about 6 times higher than the risk of developing this condition from the vaccine [161]. Further, Johnson & Johnson's Janssen (J&J/Janssen) (Ad26.COV2.S) vaccine, which was the third most commonly used vaccine early on in the pandemic, utilized an adenoviral vector to encode and display the SARS-CoV-2 S protein on a replication-deficient virus particle, a technology that has previously been used for the Ebola Virus Vaccine [162], the Human Immunodeficiency Virus (HIV) Vaccine, and the Malaria Vaccine [163,164]. Similarly, the University of Oxford/AstraZeneca (AZD1222(ChAdOx1_nCoV19) COVID-19 vaccine used an adenovirus vector producing the SARS-CoV-2 S protein [165]. However, these vaccines were associated with a risk of thrombocytopenia syndrome (TTS) resulting in blood clots. After a total of ~14 million people received the J&J vaccine in the US, and the Oxford/AstraZeneca vaccine underwent phase 1, 2, and 3 clinical trials, TTS was reported in 3.23 per million doses of J&J vaccine administered and in 7.5 per million doses of Oxford/AstraZeneca vaccine administered, with ~20–40% of TTS cases leading to death [166]. While the incidence of TTS was very rare, particularly compared to the case fatality rate of COVID-19 (0.1–5% depending on country), the cause underlying the development of TTS remains unknown and distribution of this type of vaccine in the US was halted in 2023. The mRNA vaccines, on the other hand, had a high incidence of mild side effects and a low incidence of more serious but temporary and treatable side effects. Importantly, no deaths have been attributed to the mRNA COVID vaccines. In addition, Guillaine–Barre syndrome, which is a rare auto-immune condition that sometimes develops after any vaccinations, was found to be 21 time lower after mRNA vaccinations compared to adenoviral vector vaccines against COVID-19 [32]. Recent studies have begun to investigate the molecular mechanisms underlying myocarditis pathogenesis after the SARS-CoV-2 S protein mRNA vaccines and found no difference in immune response but an increase in circulating full-length spike protein, indicating the possibility that the spike protein may have a biologic effect [167,168]. Further research elucidating the pathogenesis mechanism may lead to our ability to design mutated biologically inactive spike protein antigens, either as protein vaccines or as mRNA vaccines.

### 6.2. Developments in mRNA Immunotherapy to Botulinum Neurotoxins

Even before the mRNA vaccines developed during the COVID-19 pandemic, this technology was examined as a countermeasure for botulism using a passive vaccine strategy. Two elegant studies have shown that mRNA vaccines encoding antibodies or antibody fragments can be used to protect against botulism after toxin exposure. In this strategy, host cells take up the nanoparticle-coated mRNAs and antibody fragments are rapidly produced by the host cells and released into the circulatory system to neutralize botulinum neurotoxins before neuronal cell entry [169,170]. One of the studies explored the use of chemically unmodified mRNA to produce neutralizing antibody derivatives [169]. This study designed mRNA encoding heavy chain only VH-based neutralizing agents (VNAs). VNAs have significant biophysical advantages compared to traditional antibodies due to the small size of the VH-domains (14 kDa), ease of production, improved stability, and tissue penetration characteristics. A single injection of mRNA-lipid nanoparticle (LNP) resulted in rapid and long-lasting serum VNA titers, detectable 2 h post injection, providing prophylactic and therapeutic protection against botulinum toxin intoxication [169]. Overall, this study demonstrated the effectiveness of an mRNA vaccine as a potent technology for passive immunization. In a second study, the same group employed mRNA encoding linked VH-domains to create VNAs that recognize multiple serotypes of BoNTs [170]. This study created multiple heterohexameric VNA antitoxins composed of six camelid sdAb (single domain antibody) components (VHs), with two capable of neutralizing BoNT serotypes, A, B, or E, respectively. To test the potential of mRNA therapeutics encoding long sdAb heteromultimers, a heterohexamer was encoded as replicating RNA (repRNA), formulated with a cationic nanocarrier, and delivered to mice via intramuscular injection. Heterohexamer antitoxin serum expression levels were detectable by 8 h post treatment, peaked at 5–10 nM around two days, and remained for more than three days [170]. The heterohexamer constructs protected mice from at least 100 MIPLD50, (mouse intraperitoneal median lethal dose) of each serotype [170]. This study demonstrated the potential of long sdAb multimers, which can be administered as proteins or repRNA, in the development of antibody-based therapeutics [170].

Overall, these developments indicate a promising approach that may be extended to protective passive vaccination strategies for other toxins or organisms in the future, in particular, if safety barriers based on protein studies are included to ensure expressed antigens will be atoxic in the host. The studies described here used knowledge gained from monoclonal antibody (mAb) studies to select optimal neutralizing antibodies [171]. The mAbs-based therapies are expanding, but mAB production is expensive and not easily scaled to high output levels. Utilizing this mRNA approach provides a less expensive and easy-to-scale alternative, although there is a delay of several hours between the administration of the mRNA and the production of the neutralizing antibodies. Can a similar mRNA approach be used to create a protective anti-BoNT vaccine by encoding the toxin? In addition to hurdles due to the large size of the protein toxin, which would require the construction of a large mRNA, the idea of using mRNA to produce toxins inside the host cells is negated by the toxicity of BoNTs. However, biologically inactive subunits or genetically inactivated holotoxins may be good candidates for such developments. Detailed structure-function studies, toxicity analyses, and pharmacokinetics on the protein level will be necessary to ensure the safety and efficacy of such an approach.

## 7. Discussion and Future Directions

Vaccines have been one of the greatest contributors to improved population health. The recurring question of whether vaccines are safe for use often arises, in particular when new technologies are introduced. While vaccine safety varies based on the vaccine technology used as well as the pathogen, our medical history has shown that vaccines are always safer than the diseases the vaccines are meant to prevent or lessen in severity, and there has been a steady progression in increased safety over time.

Whole organism vaccines have the theoretical and, in some cases, demonstrated the possibility of the vaccine strain reverting to a pathogenic strain or being incompletely inactivated [22]. Recombinant protein or subunit vaccines, on the other hand, completely eliminate this risk as they do not contain whole organisms and rely on host immune responses against identified immunodominant antigens from bacteria, viruses, or toxins [8].

Also, since these vaccine formulations consist of recombinant protein specifically, protective immunity against infection is achieved without potential risks associated with other parts of the pathogen [9,172]. Recombinant and toxoid vaccines are comparatively favorable in terms of production, transportation, and storage. The vaccine candidate can be lyophilized to powder form, has a better shelf life, and can be transported and stored at room temperature as well [173]. On the other hand, recombinant and toxoid vaccines are more costly to develop and usually require the addition of adjuvant as part of vaccine formulation to achieve effective immune responses, and inducing a protective T-cell response and immune memory in addition to the humoral response is challenging. Recombinant protein vaccines can be replaced by viral vector vaccines, which in general have been clinically successful [174]; however, questions such as the use of viral vector backbone and its impact on immune response, in particular with boosters vaccinations, remain unanswered [175,176]. Viral vector vaccines are inappropriate for toxin vaccines, as they would result in the production of toxins in the host and the potential incorporation of toxin-encoding DNA into the host genome. Utilizing specific mutations in the toxin gene to knock out toxicity, as is currently being explored for recombinant toxoid vaccines, would mitigate the risk of active toxin production, but still, a risk of DNA incorporation and reversion remains. This risk is mitigated by utilizing an mRNA approach, which results in a short burst of (mutated and inactivated) protein expression followed by the destruction of the mRNA.

Interestingly, both mRNA and its specific delivery systems provide a self-adjuvanting effect resulting in long-lived and robust adaptive immune responses in mRNA-based vaccines [177]. As mRNA theoretically can encode any antigen, they can be designed against numerous targets and toxin variants with minimal formulation and adaptation to antigen alteration [178], with rapid manufacture. However, mRNA vaccines require specialized storage and transport conditions, including ultra-cold temperatures, which can be challenging and expensive. This characteristic becomes much more significant when the vaccine technology is required to be transferred to low- and middle-income countries, where limited resources are available to manufacture and store mRNA-based vaccines (storage and transportation of the vaccine vials at (−70 °C/−94 °F) are not easily accessible) [179–181].

All recombinant and toxoid vaccines share the necessity of booster shots at regular intervals to generate a strong and effective immune response, and particularly to create immune memory [18]. Table 1 provides an overview of the vaccine technologies discussed in this manuscript and their advantages and disadvantages.

Although there have been remarkable advancements in vaccine development in recent years, it is important to recognize that not all vaccines can generate a defensive immune response in the host or confer lifelong protection to all recipients. To improve existing vaccines, new strategies are required to address unmet needs, particularly for pathogens with antigenic hypervariability (such as human immunodeficiency virus) or pathogens with multiple serotypes such as dengue viruses and *S. pneumoniae*, and for pathogenic exotoxins [9]. Advances in areas such as recombinant DNA technology, bioinformatics, molecular biology, and immunology have provided helpful insights into understanding the interaction of pathogen cells and toxins with host cells. These advancements have the potential to lead to the development of multiple new vaccine strategies by specifically targeting subunits or toxin domains involved in host pathogenesis. However, in conjunction with discovering safe and effective antigens, adjuvants, and delivery systems, the challenges of vaccine development extend to evaluating the balance between cost, benefits, and risks before translating a vaccine candidate to the clinic. The success of vaccines in the future relies on determining which technology, alone or in combination, is optimal for each pathogen or toxin. This answer again is dependent on the pathogen or toxin properties and mechanisms leading to pathogenesis in the host. Overall, much work remains to be carried out at the forefront of vaccine developments, to optimize vaccine candidates, production strategies, and immune-modulating adjuvants. Much of what is learned from vaccines targeting whole pathogens or subunits can be applied to the design of improved toxin vaccines, and visa versa. For botulinum and tetanus toxin vaccines, recent research determined that the entire toxin protein results in the most robust immune response, and that it is possible to introduce relatively few targeted amino acid substitutions to delete the biologic activity of the toxin while maintaining overall structure and antigenic epitopes. The same idea of selectively mutating single amino acid residues to create a vaccine consisting of a biologically null recombinant protein can also be utilized to improve the safety profile of recombinant subunit vaccines, some of which may result in undesired side effects in specific subpopulations. Finally, this approach is also easily translated towards mRNA vaccines, where the mRNA code is easily altered to produce a biologically inactive but highly immunogenic protein. Most promising perhaps is the combination of toxicity research using recombinant toxins and subunit proteins, and the mRNA technology for vaccine delivery. Overall, recent progress in vaccine technology and our understanding of molecular mechanisms involved in pathogenesis offer a promising and bright future for new vaccine developments.

**Table 1.** Advantages and disadvantages corresponding to different vaccine types.

| Vaccine Type | Advantages | Disadvantages |
| --- | --- | --- |
| Live vaccines (attenuated) | ▪ Elicit humoral as well as cellular immune responses.<br>▪ Long-lasting protection<br>▪ Single dose is normally sufficient. | ▪ Risk of reversion to pathogenic strain.<br>▪ Complex attenuation process required.<br>▪ Vaccine composition not well defined. |
| Killed vaccines | ▪ The vaccine strain does not revert back to pathogenic strain.<br>▪ Usually only mild side effects such as fever and nausea observed.<br>▪ Inexpensive to produce at mass level. | ▪ Multiple boosters required.<br>▪ Addition of adjuvants required.<br>▪ Vaccine elicits primarily humoral responses.<br>▪ Vaccine composition not well defined.<br>▪ Large amounts of live pathogens must be cultured to produce vaccine antigens. |
| Toxoids | ▪ Toxicity inactivated.<br>▪ Low incidence of mild adverse events such as soreness or swelling, fever, nausea.<br>▪ Good stability.<br>▪ Doesn't require cold storage. | ▪ Multiple doses required to achieve immune memory.<br>▪ Addition of adjuvants required.<br>▪ Local reactions to residual formaldehyde used for inactivation.<br>▪ Requires large-scale production of active toxins.<br>▪ Immune epitopes after inactivation procedure not well defined. |
| Subunit vaccines | ▪ Composition well defined.<br>▪ Can be formulated with various vaccine delivery systems such as adjuvant, viral vector, nanoparticles, etc.<br>▪ Live pathogen is not used to produce the antigen.<br>▪ The vaccine strain cannot revert back to pathogenic strain. | ▪ Multiple doses or boosters required for immune memory.<br>▪ Antibody response restricted to subunit present in vaccine.<br>▪ Addition of adjuvants required.<br>▪ Live pathogens may be cultured to purify the vaccine antigen. |
| Recombinant toxin vaccines | ▪ Can utilize atoxic toxin sub-domains.<br>▪ Can inactivate toxicity by specific mutations without altering immunogenic surface epitopes.<br>▪ Safe to produce as no active toxin is involved. | ▪ Multiple doses or boosters required.<br>▪ Addition of adjuvants required.<br>▪ More expensive and complex to develop compared to other vaccine types. |
| mRNA vaccines | ▪ Can be quickly designed, tested, and mass produced.<br>▪ Do not consist of live pathogens.<br>▪ Large number of uses during COVID pandemic indicates good safety and effectiveness. | ▪ Requires storage in extreme cold conditions.<br>▪ Multiple doses or boosters required.<br>▪ More data on safety are required.<br>▪ Potential of adverse side effects caused by produced antigen. |

**Author Contributions:** Conceptualization, S.G. and S.P.; methodology, S.G. and S.P.; software, S.G.; validation, S.G. and S.P.; formal analysis, S.G. and S.P.; investigation, S.G.; resources, S.P.; data curation, S.G. amd S.P.; writing—original draft preparation, S.G.; writing—review and editing, S.G. and S.P.; visualization, S.G.; supervision, S.P.; project administration, S.P.; funding acquisition, S.P. All authors have read and agreed to the published version of the manuscript.

**Funding:** Funding for this work was received from the National Institutes of Health (NIH) Grant # 2R01AI118389.

**Institutional Review Board Statement:** Not applicable.

**Informed Consent Statement:** Not applicable.

**Data Availability Statement:** Not applicable.

**Conflicts of Interest:** The authors declare no conflict of interest. S.P. is a co-inventor on patent-US011491239B2.

## References

1. Melief, C.J.; van Hall, T.; Arens, R.; Ossendorp, F.; van der Burg, S.H. Therapeutic cancer vaccines. *J. Clin. Investig.* **2015**, *125*, 3401–3412. [CrossRef]
2. Bol, K.F.; Aarntzen, E.H.; Pots, J.M.; Olde Nordkamp, M.A.; van de Rakt, M.W.; Scharenborg, N.M.; de Boer, A.J.; van Oorschot, T.G.; Croockewit, S.A.; Blokx, W.A.; et al. Prophylactic vaccines are potent activators of monocyte-derived dendritic cells and drive effective anti-tumor responses in melanoma patients at the cost of toxicity. *Cancer Immunol. Immunother.* **2016**, *65*, 327–339. [CrossRef] [PubMed]
3. Riedel, S. Edward Jenner and the history of smallpox and vaccination. In *Baylor University Medical Center Proceedings*; Taylor & Francis: Oxford, UK, 2005; Volume 18, pp. 21–25. [CrossRef]
4. Plotkin, S. History of vaccination. *Proc. Natl. Acad. Sci. USA* **2014**, *111*, 12283–12287. [CrossRef] [PubMed]
5. Nabel, G.J. Designing tomorrow's vaccines. *N. Engl. J. Med.* **2013**, *368*, 551–560. [CrossRef]
6. Plotkin, S.A. Vaccination in the 21st century. *J. Infect. Dis.* **1993**, *168*, 29–37. [CrossRef]
7. Plotkin, S.A. Vaccines: Past, present and future. *Nat. Med.* **2005**, *11*, S5–S11. [CrossRef]
8. Clark, T.G.; Cassidy-Hanley, D. Recombinant subunit vaccines: Potentials and constraints. *Dev. Biol.* **2005**, *121*, 153–163.
9. Nascimento, I.P.; Leite, L.C. Recombinant vaccines and the development of new vaccine strategies. *Braz. J. Med. Biol. Res.* **2012**, *45*, 1102–1111. [CrossRef] [PubMed]
10. Simborio, H.L.; Lee, J.J.; Bernardo Reyes, A.W.; Hop, H.T.; Arayan, L.T.; Min, W.; Lee, H.J.; Yoo, H.S.; Kim, S. Evaluation of the combined use of the recombinant Brucella abortus Omp10, Omp19 and Omp28 proteins for the clinical diagnosis of bovine brucellosis. *Microb. Pathog.* **2015**, *83–84*, 41–46. [CrossRef]
11. Hotez, P.J.; Bottazzi, M.E. Whole Inactivated Virus and Protein-Based COVID-19 Vaccines. *Annu. Rev. Med.* **2022**, *73*, 55–64. [CrossRef]
12. Lehrer, A.T.; Chuang, E.; Namekar, M.; Williams, C.A.; Wong, T.A.S.; Lieberman, M.M.; Granados, A.; Misamore, J.; Yalley-Ogunro, J.; Andersen, H.; et al. Recombinant Protein Filovirus Vaccines Protect Cynomolgus Macaques from Ebola, Sudan, and Marburg Viruses. *Front. Immunol.* **2021**, *12*, 703986. [CrossRef]
13. Prygiel, M.; Mosiej, E.; Gorska, P.; Zasada, A.A. Diphtheria-tetanus-pertussis vaccine: Past, current & future. *Future Microbiol.* **2022**, *17*, 185–197. [CrossRef] [PubMed]
14. Smith, J.W. Diphtheria and tetanus toxoids. *Br. Med. Bull.* **1969**, *25*, 177–182. [CrossRef]
15. Przedpelski, A.; Tepp, W.H.; Pellett, S.; Johnson, E.A.; Barbieri, J.T. A Novel High-Potency Tetanus Vaccine. *mBio* **2020**, *11*, 01668-20. [CrossRef] [PubMed]
16. Graham, R.; Thorp, F. The Effect of Formalin on Botulinum Toxins A, B and C. *J. Immunol.* **1929**, *16*, 391–401. [CrossRef]
17. Aquino, M.R.; Bingemann, T.A.; Nanda, A.; Maples, K.M. Delayed allergic skin reactions to vaccines. *Allergy Asthma Proc.* **2022**, *43*, 20–29. [CrossRef] [PubMed]
18. Verbeke, R.; Lentacker, I.; De Smedt, S.C.; Dewitte, H. The dawn of mRNA vaccines: The COVID-19 case. *J. Control. Release* **2021**, *333*, 511–520. [CrossRef]
19. Jain, S.; Venkataraman, A.; Wechsler, M.E.; Peppas, N.A. Messenger RNA-based vaccines: Past, present, and future directions in the context of the COVID-19 pandemic. *Adv. Drug. Deliv. Rev.* **2021**, *179*, 114000. [CrossRef]
20. Delany, I.; Rappuoli, R.; De Gregorio, E. Vaccines for the 21st century. *EMBO Mol. Med.* **2014**, *6*, 708–720. [CrossRef]
21. Akkaya, M.; Kwak, K.; Pierce, S.K. B cell memory: Building two walls of protection against pathogens. *Nat. Rev. Immunol.* **2020**, *20*, 229–238. [CrossRef] [PubMed]
22. Minor, P.D. Live attenuated vaccines: Historical successes and current challenges. *Virology* **2015**, *479–480*, 379–392. [CrossRef]
23. Strebel, P.M.; Papania, M.J.; Fiebelkorn, A.P.; Halsey, N.A. 20—Measles vaccine. In *Vaccines*, 6th ed.; Plotkin, S.A., Orenstein, W.A., Offit, P.A., Eds.; W.B. Saunders: London, UK, 2013; pp. 352–387. [CrossRef]

24. Lieberman, J.M.; Williams, W.R.; Miller, J.M.; Black, S.; Shinefield, H.; Henderson, F.; Marchant, C.D.; Werzberger, A.; Halperin, S.; Hartzel, J.; et al. The safety and immunogenicity of a quadrivalent measles, mumps, rubella and varicella vaccine in healthy children: A study of manufacturing consistency and persistence of antibody. *Pediatr. Infect. Dis. J.* **2006**, *25*, 615–622. [CrossRef] [PubMed]
25. Patja, A.; Davidkin, I.; Kurki, T.; Kallio, M.J.; Valle, M.; Peltola, H. Serious adverse events after measles-mumps-rubella vaccination during a fourteen-year prospective follow-up. *Pediatr. Infect. Dis. J.* **2000**, *19*, 1127–1134. [CrossRef] [PubMed]
26. Pace, J.L.; Rossi, H.A.; Esposito, V.M.; Frey, S.M.; Tucker, K.D.; Walker, R.I. Inactivated whole-cell bacterial vaccines: Current status and novel strategies. *Vaccine* **1998**, *16*, 1563–1574. [CrossRef]
27. Bandyopadhyay, A.S.; Garon, J.; Seib, K.; Orenstein, W.A. Polio vaccination: Past, present and future. *Future Microbiol.* **2015**, *10*, 791–808. [CrossRef]
28. Hird, T.R.; Grassly, N.C. Systematic review of mucosal immunity induced by oral and inactivated poliovirus vaccines against virus shedding following oral poliovirus challenge. *PLoS Pathog.* **2012**, *8*, e1002599. [CrossRef]
29. World Health Organization. Supplementary Information on Vaccine Safety. Part 2: Background Rates of Adverse Events Following Immunization. World Health Organization. 2000. Available online: https://apps.who.int/iris/handle/10665/66675 (accessed on 1 July 2023).
30. Ren, Z.; Sun, R.; Cui, G.; Wang, H.; Zhang, D.; Li, J.; Zhang, Y.; Yu, Z. Effects of Inactivated Vaccination on Humoral Immune Responses in Patients Infected with Delta or Omicron Variants. *J. Infect. Dis.* **2022**, *226*, 1120–1122. [CrossRef]
31. Excler, J.L.; Saville, M.; Berkley, S.; Kim, J.H. Vaccine development for emerging infectious diseases. *Nat. Med.* **2021**, *27*, 591–600. [CrossRef]
32. CDC. Adverse Reaction. 2022. Available online: https://www.cdc.gov/rabies/specific_groups/hcp/adverse_reaction.html#:~:text=Immediate%20hypersensitivity%20reactions%20have%20been,6%2D14%20days%20after%20boosters (accessed on 1 July 2023).
33. Liang, J.L.; Tiwari, T.; Moro, P.; Messonnier, N.E.; Reingold, A.; Sawyer, M.; Clark, T.A. Prevention of Pertussis, Tetanus, and Diphtheria with Vaccines in the United States: Recommendations of the Advisory Committee on Immunization Practices (ACIP). *MMWR Recomm. Rep.* **2018**, *67*, 1–44. [CrossRef]
34. McQuillan, G.M.; Kruszon-Moran, D.; Deforest, A.; Chu, S.Y.; Wharton, M. Serologic immunity to diphtheria and tetanus in the United States. *Ann. Intern. Med.* **2002**, *136*, 660–666. [CrossRef] [PubMed]
35. Angsantikul, P.; Fang, R.H.; Zhang, L. Toxoid Vaccination against Bacterial Infection Using Cell Membrane-Coated Nanoparticles. *Bioconjug. Chem.* **2018**, *29*, 604–612. [CrossRef] [PubMed]
36. Shaw, C.A.; Starnbach, M.N. Stimulation of CD8 + T cells following diphtheria toxin-mediated antigen delivery into dendritic cells. *Infect. Immun.* **2006**, *74*, 1001–1008. [CrossRef]
37. Skibinski, D.A.; Baudner, B.C.; Singh, M.; O'Hagan, D.T. Combination vaccines. *J. Glob. Infect. Dis.* **2011**, *3*, 63–72. [CrossRef]
38. Farrar, J.; Newton, C. Neurological aspects of tropical disease. *J. Neurol. Neurosurg. Psychiatry* **2000**, *68*, 135–136. [CrossRef] [PubMed]
39. Stock, I. Tetanus and Clostridium tetani—A brief review. *Med. Monatsschr. Pharm.* **2015**, *38*, 57–60.
40. Rhinesmith, E.; Fu, L. Tetanus Disease, Treatment, Management. *Pediatr. Rev.* **2018**, *39*, 430–432. [CrossRef]
41. Goretzki, K.; Habermann, E. Enzymatic hydrolysis of tetanus toxin by intrinsic and extrinsic proteases. Characterization of the fragments by monoclonal antibodies. *Med. Microbiol. Immunol.* **1985**, *174*, 139–150. [CrossRef] [PubMed]
42. Farrar, J.J.; Yen, L.M.; Cook, T.; Fairweather, N.; Binh, N.; Parry, J.; Parry, C.M. Tetanus. *J. Neurol. Neurosurg. Psychiatry* **2000**, *69*, 292–301. [CrossRef]
43. Behrensdorf-Nicol, H.A.; Kegel, B.; Bonifas, U.; Silberbach, K.; Klimek, J.; Weiber, K.; Kramer, B. Residual enzymatic activity of the tetanus toxin light chain present in tetanus toxoid batches used for vaccine production. *Vaccine* **2008**, *26*, 3835–3841. [CrossRef]
44. Stojicevic, I.; Dimitrijevic, L.; Dovezenski, N.; Zivkovic, I.; Petrusic, V.; Marinkovic, E.; Inic-Kanada, A.; Stojanovic, M. Tetanus toxoid purification: Chromatographic procedures as an alternative to ammonium-sulphate precipitation. *J. Chromatogr. B Anal. Technol. Biomed. Life Sci.* **2011**, *879*, 2213–2219. [CrossRef]
45. Rossetto, O.; Montecucco, C. Tables of Toxicity of Botulinum and Tetanus Neurotoxins. *Toxins* **2019**, *11*, 686. [CrossRef] [PubMed]
46. Thaysen-Andersen, M.; Jorgensen, S.B.; Wilhelmsen, E.S.; Petersen, J.W.; Hojrup, P. Investigation of the detoxification mechanism of formaldehyde-treated tetanus toxin. *Vaccine* **2007**, *25*, 2213–2227. [CrossRef] [PubMed]
47. Michiels, T.J.M.; Meiring, H.D.; Jiskoot, W.; Kersten, G.F.A.; Metz, B. Formaldehyde treatment of proteins enhances proteolytic degradation by the endo-lysosomal protease cathepsin S. *Sci. Rep.* **2020**, *10*, 11535. [CrossRef]
48. Fan, Y.C.; Chiu, H.C.; Chen, L.K.; Chang, G.J.; Chiou, S.S. Formalin Inactivation of Japanese Encephalitis Virus Vaccine Alters the Antigenicity and Immunogenicity of a Neutralization Epitope in Envelope Protein Domain III. *PLoS Neglected Trop. Dis.* **2015**, *9*, e0004167. [CrossRef]
49. Metz, B.; Michiels, T.; Uittenbogaard, J.; Danial, M.; Tilstra, W.; Meiring, H.D.; Hennink, W.E.; Crommelin, D.J.A.; Kersten, G.F.A.; Jiskoot, W. Identification of Formaldehyde-Induced Modifications in Diphtheria Toxin. *J. Pharm. Sci.* **2020**, *109*, 543–557. [CrossRef]
50. Michiels, T.J.M.; Schoneich, C.; Hamzink, M.R.J.; Meiring, H.D.; Kersten, G.F.A.; Jiskoot, W.; Metz, B. Novel Formaldehyde-Induced Modifications of Lysine Residue Pairs in Peptides and Proteins: Identification and Relevance to Vaccine Development. *Mol. Pharm.* **2020**, *17*, 4375–4385. [CrossRef]

51. Mc, C.J.; Trafton, M.Z. Immune responses and reactions to diphtheria and tetanus toxoids, with pertussis vaccine, aluminum phosphate precipitated. *N. Engl. J. Med.* **1950**, *243*, 442–444. [CrossRef]
52. Ghotloo, S.; Golsaz-Shirazi, F.; Amiri, M.M.; Jeddi-Tehrani, M.; Shokri, F. Epitope Mapping of Tetanus Toxin by Monoclonal Antibodies: Implication for Immunotherapy and Vaccine Design. *Neurotox. Res.* **2020**, *37*, 239–249. [CrossRef] [PubMed]
53. Blencowe, H.; Lawn, J.; Vandelaer, J.; Roper, M.; Cousens, S. Tetanus toxoid immunization to reduce mortality from neonatal tetanus. *Int. J. Epidemiol.* **2010**, *39* (Suppl. S1), i102–i109. [CrossRef]
54. Kyu, H.H.; Mumford, J.E.; Stanaway, J.D.; Barber, R.M.; Hancock, J.R.; Vos, T.; Murray, C.J.; Naghavi, M. Mortality from tetanus between 1990 and 2015: Findings from the global burden of disease study 2015. *BMC Public Health* **2017**, *17*, 179. [CrossRef] [PubMed]
55. Sharma, N.C.; Efstratiou, A.; Mokrousov, I.; Mutreja, A.; Das, B.; Ramamurthy, T. Diphtheria. *Nat. Rev. Dis. Prim.* **2019**, *5*, 81. [CrossRef]
56. Holbourn, K.P.; Shone, C.C.; Acharya, K.R. A family of killer toxins. Exploring the mechanism of ADP-ribosylating toxins. *FEBS J.* **2006**, *273*, 4579–4593. [CrossRef] [PubMed]
57. World Health Organization. Diptheria Reported Cases by WHO Region. In *WHO Site*; World Health Organization: Geneva, Switzerland, 2023.
58. Gill, D.M. Bacterial toxins: A table of lethal amounts. *Microbiol. Rev.* **1982**, *46*, 86–94. [CrossRef] [PubMed]
59. Hill, K.K.; Smith, T.J.; Helma, C.H.; Ticknor, L.O.; Foley, B.T.; Svensson, R.T.; Brown, J.L.; Johnson, E.A.; Smith, L.A.; Okinaka, R.T.; et al. Genetic diversity among Botulinum Neurotoxin-producing clostridial strains. *J. Bacteriol.* **2007**, *189*, 818–832. [CrossRef]
60. Schiavo, G.; Matteoli, M.; Montecucco, C. Neurotoxins affecting neuroexocytosis. *Physiol. Rev.* **2000**, *80*, 717–766. [CrossRef]
61. Montecucco, C.; Rossetto, O.; Schiavo, G. Presynaptic receptor arrays for clostridial neurotoxins. *Trends Microbiol.* **2004**, *12*, 442–446. [CrossRef]
62. Lam, K.H.; Yao, G.; Jin, R. Diverse binding modes, same goal: The receptor recognition mechanism of botulinum neurotoxin. *Prog. Biophys. Mol. Biol.* **2015**, *117*, 225–231. [CrossRef]
63. Lacy, D.B.; Tepp, W.; Cohen, A.C.; DasGupta, B.R.; Stevens, R.C. Crystal structure of botulinum neurotoxin type A and implications for toxicity. *Nat. Struct. Biol.* **1998**, *5*, 898–902. [CrossRef]
64. Whitemarsh, R.C.; Tepp, W.H.; Johnson, E.A.; Pellett, S. Persistence of botulinum neurotoxin a subtypes 1–5 in primary rat spinal cord cells. *PLoS ONE* **2014**, *9*, e90252. [CrossRef]
65. Pellett, S.; Bradshaw, M.; Tepp, W.H.; Pier, C.L.; Whitemarsh, R.C.M.; Chen, C.; Barbieri, J.T.; Johnson, E.A. The Light Chain Defines the Duration of Action of Botulinum Toxin Serotype A Subtypes. *mBio* **2018**, *9*, 00089-18. [CrossRef] [PubMed]
66. Gardner, A.P.; Barbieri, J.T. Light Chain Diversity among the Botulinum Neurotoxins. *Toxins* **2018**, *10*, 268. [CrossRef] [PubMed]
67. Hatheway, C.L. Botulism: The present status of the disease. *Curr. Top. Microbiol. Immunol.* **1995**, *195*, 55–75. [CrossRef] [PubMed]
68. Carrillo-Marquez, M.A. Botulism. *Pediatr. Rev.* **2016**, *37*, 183–192. [CrossRef] [PubMed]
69. Sundeen, G.; Barbieri, J.T. Vaccines against Botulism. *Toxins* **2017**, *9*, 268. [CrossRef]
70. Karalewitz, A.P.; Barbieri, J.T. Vaccines against botulism. *Curr. Opin. Microbiol.* **2012**, *15*, 317–324. [CrossRef]
71. Centers for Disease Control and Prevention. Notice of CDC's discontinuation of investigational pentavalent (ABCDE) botulinum toxoid vaccine for workers at risk for occupational exposure to botulinum toxins. *MMWR Morb. Mortal. Wkly. Rep.* **2011**, *60*, 1454–1455.
72. Rusnak, J.M.; Smith, L.A. Botulinum neurotoxin vaccines: Past history and recent developments. *Hum. Vaccin.* **2009**, *5*, 794–805. [CrossRef]
73. Smith, L.A.; Rusnak, J.M. Botulinum neurotoxin vaccines: Past, present, and future. *Crit. Rev. Immunol.* **2007**, *27*, 303–318. [CrossRef]
74. Pier, C.L.; Tepp, W.H.; Bradshaw, M.; Johnson, E.A.; Barbieri, J.T.; Baldwin, M.R. Recombinant holotoxoid vaccine against botulism. *Infect. Immun.* **2008**, *76*, 437–442. [CrossRef]
75. Bradshaw, M.; Goodnough, M.C.; Johnson, E.A. Conjugative transfer of the Escherichia coli-Clostridium perfringens shuttle vector pJIR1457 to Clostridium botulinum type A strains. *Plasmid* **1998**, *40*, 233–237. [CrossRef]
76. Johnson, E.A. Clostridial toxins as therapeutic agents: Benefits of nature's most toxic proteins. *Annu. Rev. Microbiol.* **1999**, *53*, 551–575. [CrossRef] [PubMed]
77. Whitemarsh, R.C.; Tepp, W.H.; Bradshaw, M.; Lin, G.; Pier, C.L.; Scherf, J.M.; Johnson, E.A.; Pellett, S. Characterization of botulinum neurotoxin A subtypes 1 through 5 by investigation of activities in mice, in neuronal cell cultures, and in vitro. *Infect. Immun.* **2013**, *81*, 3894–3902. [CrossRef] [PubMed]
78. Smith, L.A. Development of recombinant vaccines for botulinum neurotoxin. *Toxicon* **1998**, *36*, 1539–1548. [CrossRef]
79. Torii, Y.; Sugimoto, N.; Kohda, T.; Kozaki, S.; Morokuma, K.; Horikawa, Y.; Ginnaga, A.; Yamamoto, A.; Takahashi, M. Clinical Study of New Tetravalent (Type A, B, E, and F) Botulinum Toxoid Vaccine Derived from M Toxin in Japan. *Jpn. J. Infect. Dis.* **2017**, *70*, 423–429. [CrossRef] [PubMed]
80. Kobayashi, R.; Kohda, T.; Kataoka, K.; Ihara, H.; Kozaki, S.; Pascual, D.W.; Staats, H.F.; Kiyono, H.; McGhee, J.R.; Fujihashi, K. A novel neurotoxoid vaccine prevents mucosal botulism. *J. Immunol.* **2005**, *174*, 2190–2195. [CrossRef]
81. Torii, Y.; Tokumaru, Y.; Kawaguchi, S.; Izumi, N.; Maruyama, S.; Mukamoto, M.; Kozaki, S.; Takahashi, M. Production and immunogenic efficacy of botulinum tetravalent (A, B, E, F) toxoid. *Vaccine* **2002**, *20*, 2556–2561. [CrossRef]
82. Dertzbaugh, M.T. Genetically engineered vaccines: An overview. *Plasmid* **1998**, *39*, 100–113. [CrossRef]

83. Greenberg, L. A New Approach to Bacterial Vaccines. *Can. Med. Assoc. J.* **1963**, *89*, 396–402.
84. Sivakumar, S.M.; Safhi, M.M.; Kannadasan, M.; Sukumaran, N. Vaccine adjuvants—Current status and prospects on controlled release adjuvancity. *Saudi. Pharm. J.* **2011**, *19*, 197–206. [CrossRef]
85. Gupta, R.K.; Rost, B.E.; Relyveld, E.; Siber, G.R. Adjuvant properties of aluminum and calcium compounds. *Pharm. Biotechnol.* **1995**, *6*, 229–248. [CrossRef] [PubMed]
86. Diaz-Dinamarca, D.A.; Salazar, M.L.; Castillo, B.N.; Manubens, A.; Vasquez, A.E.; Salazar, F.; Becker, M.I. Protein-Based Adjuvants for Vaccines as Immunomodulators of the Innate and Adaptive Immune Response: Current Knowledge, Challenges, and Future Opportunities. *Pharmaceutics* **2022**, *14*, 1671. [CrossRef] [PubMed]
87. Pulendran, B.; Arunachalam, P.S.; O'Hagan, D.T. Emerging concepts in the science of vaccine adjuvants. *Nat. Rev. Drug Discov.* **2021**, *20*, 454–475. [CrossRef] [PubMed]
88. Verma, S.K.; Mahajan, P.; Singh, N.K.; Gupta, A.; Aggarwal, R.; Rappuoli, R.; Johri, A.K. New-age vaccine adjuvants, their development, and future perspective. *Front. Immunol.* **2023**, *14*, 1043109. [CrossRef]
89. Slifka, M.K.; Amanna, I. How advances in immunology provide insight into improving vaccine efficacy. *Vaccine* **2014**, *32*, 2948–2957. [CrossRef]
90. Coffman, R.L.; Sher, A.; Seder, R.A. Vaccine adjuvants: Putting innate immunity to work. *Immunity* **2010**, *33*, 492–503. [CrossRef]
91. Schwarz, T.F.; Horacek, T.; Knuf, M.; Damman, H.G.; Roman, F.; Drame, M.; Gillard, P.; Jilg, W. Single dose vaccination with AS03-adjuvanted H5N1 vaccines in a randomized trial induces strong and broad immune responsiveness to booster vaccination in adults. *Vaccine* **2009**, *27*, 6284–6290. [CrossRef]
92. Zhao, H.; Zhou, X.; Zhou, Y.H. Hepatitis B vaccine development and implementation. *Hum. Vaccine Immunother.* **2020**, *16*, 1533–1544. [CrossRef]
93. Adkins, J.C.; Wagstaff, A.J. Recombinant hepatitis B vaccine: A review of its immunogenicity and protective efficacy against hepatitis B. *BioDrugs* **1998**, *10*, 137–158. [CrossRef]
94. Govan, V.A. A novel vaccine for cervical cancer: Quadrivalent human papillomavirus (types 6, 11, 16 and 18) recombinant vaccine (Gardasil). *Ther. Clin. Risk Manag.* **2008**, *4*, 65–70. [CrossRef]
95. Shinefield, H.R. Overview of the development and current use of CRM(197) conjugate vaccines for pediatric use. *Vaccine* **2010**, *28*, 4335–4339. [CrossRef] [PubMed]
96. Rappuoli, R. Isolation and characterization of Corynebacterium diphtheriae nontandem double lysogens hyperproducing CRM197. *Appl. Environ. Microbiol.* **1983**, *46*, 560–564. [CrossRef] [PubMed]
97. Qiao, J.; Ghani, K.; Caruso, M. Diphtheria toxin mutant CRM197 is an inhibitor of protein synthesis that induces cellular toxicity. *Toxicon* **2008**, *51*, 473–477. [CrossRef] [PubMed]
98. Kageyama, T.; Ohishi, M.; Miyamoto, S.; Mizushima, H.; Iwamoto, R.; Mekada, E. Diphtheria toxin mutant CRM197 possesses weak EF2-ADP-ribosyl activity that potentiates its anti-tumorigenic activity. *J. Biochem.* **2007**, *142*, 95–104. [CrossRef] [PubMed]
99. Rabadi, T.; Brady, M.F. Tetanus Toxoid. In *StatPearls*; StatPearls Publishing: Treasure Island, FL, USA, 2023.
100. World Health Organization. *Vaccine Safety Basics*; World Health Organization: Geneva, Switzerland, 2013.
101. Liu, F.J.; Shi, D.Y.; Li, Z.Y.; Lu, J.S.; Wang, R.; Pang, X.B.; Yang, Z.X.; Yu, Y.Z. Evaluation of a recombinant tetanus toxin subunit vaccine. *Toxicon* **2020**, *187*, 75–81. [CrossRef]
102. Chang, M.J.; Ollivault-Shiflett, M.; Schuman, R.; Ngoc Nguyen, S.; Kaltashov, I.A.; Bobst, C.; Rajagopal, S.P.; Przedpelski, A.; Barbieri, J.T.; Lees, A. Genetically detoxified tetanus toxin as a vaccine and conjugate carrier protein. *Vaccine* **2022**, *40*, 5103–5113. [CrossRef]
103. Rasetti-Escargueil, C.; Popoff, M.R. Antibodies and Vaccines against Botulinum Toxins: Available Measures and Novel Approaches. *Toxins* **2019**, *11*, 528. [CrossRef]
104. LaPenotiere, H.F.; Clayton, M.A.; Middlebrook, J.L. Expression of a large, nontoxic fragment of botulinum neurotoxin serotype A and its use as an immunogen. *Toxicon* **1995**, *33*, 1383–1386. [CrossRef]
105. Webb, R.P.; Smith, L.A. What next for botulism vaccine development? *Expert Rev. Vaccines* **2013**, *12*, 481–492. [CrossRef]
106. Baldwin, M.R.; Tepp, W.H.; Przedpelski, A.; Pier, C.L.; Bradshaw, M.; Johnson, E.A.; Barbieri, J.T. Subunit vaccine against the seven serotypes of botulism. *Infect. Immun.* **2008**, *76*, 1314–1318. [CrossRef]
107. Przedpelski, A.; Tepp, W.H.; Kroken, A.R.; Fu, Z.; Kim, J.J.; Johnson, E.A.; Barbieri, J.T. Enhancing the protective immune response against botulism. *Infect. Immun.* **2013**, *81*, 2638–2644. [CrossRef]
108. Baldwin, M.R.; Tepp, W.H.; Pier, C.L.; Bradshaw, M.; Ho, M.; Wilson, B.A.; Fritz, R.B.; Johnson, E.A.; Barbieri, J.T. Characterization of the antibody response to the receptor binding domain of botulinum neurotoxin serotypes A and E. *Infect. Immun.* **2005**, *73*, 6998–7005. [CrossRef] [PubMed]
109. Khouri, J.M.; Motter, R.N.; Arnon, S.S. Safety and immunogenicity of investigational recombinant botulinum vaccine, rBV A/B, in volunteers with pre-existing botulinum toxoid immunity. *Vaccine* **2018**, *36*, 2041–2048. [CrossRef] [PubMed]
110. Webb, R.P.; Smith, T.J.; Smith, L.A.; Wright, P.M.; Guernieri, R.L.; Brown, J.L.; Skerry, J.C. Recombinant Botulinum Neurotoxin Hc Subunit (BoNT Hc) and Catalytically Inactive Clostridium botulinum Holoproteins (ciBoNT HPs) as Vaccine Candidates for the Prevention of Botulism. *Toxins* **2017**, *9*, 269. [CrossRef] [PubMed]
111. Byrne, M.P.; Smith, L.A. Development of vaccines for prevention of botulism. *Biochimie* **2000**, *82*, 955–966. [CrossRef]
112. Band, P.A.; Blais, S.; Neubert, T.A.; Cardozo, T.J.; Ichtchenko, K. Recombinant derivatives of botulinum neurotoxin A engineered for trafficking studies and neuronal delivery. *Protein Expr. Purif.* **2010**, *71*, 62–73. [CrossRef] [PubMed]

113. Tepp, W.H.; Bradshaw, M.; Gardner, A.P.; Kaufman, R.L.; Barbieri, J.T.; Pellett, S. Botulinum Neurotoxin A4 Has a 1000-Fold Reduced Potency Due to Three Single Amino Acid Alterations in the Protein Receptor Binding Domain. *Int. J. Mol. Sci.* **2023**, *24*, 5690. [CrossRef]
114. Bradshaw, M.; Tepp, W.H.; Whitemarsh, R.C.; Pellett, S.; Johnson, E.A. Holotoxin Activity of Botulinum Neurotoxin Subtype A4 Originating from a Nontoxigenic Clostridium botulinum Expression System. *Appl. Environ. Microbiol.* **2014**, *80*, 7415–7422. [CrossRef]
115. Dolly, J.O.; Wang, J.; Zurawski, T.H.; Meng, J. Novel therapeutics based on recombinant botulinum neurotoxins to normalize the release of transmitters and pain mediators. *FEBS J.* **2011**, *278*, 4454–4466. [CrossRef]
116. Rummel, A.; Mahrhold, S.; Bigalke, H.; Binz, T. Exchange of the H(CC) domain mediating double receptor recognition improves the pharmacodynamic properties of botulinum neurotoxin. *FEBS J.* **2011**, *278*, 4506–4515. [CrossRef]
117. Webb, R.P.; Smith, T.J.; Wright, P.; Brown, J.; Smith, L.A. Production of catalytically inactive BoNT/A1 holoprotein and comparison with BoNT/A1 subunit vaccines against toxin subtypes A1, A2, and A3. *Vaccine* **2009**, *27*, 4490–4497. [CrossRef]
118. Webb, R.P. Engineering of Botulinum Neurotoxins for Biomedical Applications. *Toxins* **2018**, *10*, 231. [CrossRef]
119. Smith, T.J.; Lou, J.; Geren, I.N.; Forsyth, C.M.; Tsai, R.; Laporte, S.L.; Tepp, W.H.; Bradshaw, M.; Johnson, E.A.; Smith, L.A.; et al. Sequence variation within botulinum neurotoxin serotypes impacts antibody binding and neutralization. *Infect. Immun.* **2005**, *73*, 5450–5457. [CrossRef]
120. Przedpelski, A.; Tepp, W.H.; Zuverink, M.; Johnson, E.A.; Pellet, S.; Barbieri, J.T. Enhancing toxin-based vaccines against botulism. *Vaccine* **2018**, *36*, 827–832. [CrossRef] [PubMed]
121. Li, Z.; Lu, J.; Tan, X.; Wang, R.; Xu, Q.; Yu, Y.; Yang, Z. Functional EL-HN Fragment as a Potent Candidate Vaccine for the Prevention of Botulinum Neurotoxin Serotype E. *Toxins* **2022**, *14*, 135. [CrossRef] [PubMed]
122. Li, Z.; Li, B.; Lu, J.; Liu, X.; Tan, X.; Wang, R.; Du, P.; Yu, S.; Xu, Q.; Pang, X.; et al. Biological and Immunological Characterization of a Functional L-HN Derivative of Botulinum Neurotoxin Serotype F. *Toxins* **2023**, *15*, 200. [CrossRef]
123. Ayyar, B.V.; Tajhya, R.B.; Beeton, C.; Atassi, M.Z. Antigenic sites on the HN domain of botulinum neurotoxin A stimulate protective antibody responses against active toxin. *Sci. Rep.* **2015**, *5*, 15776. [CrossRef] [PubMed]
124. Atassi, M.Z.; Dolimbek, B.Z. Mapping of the antibody-binding regions on the HN-domain (residues 449-859) of botulinum neurotoxin A with antitoxin antibodies from four host species. Full profile of the continuous antigenic regions of the H-chain of botulinum neurotoxin A. *Protein J.* **2004**, *23*, 39–52. [CrossRef]
125. Atassi, M.Z. Molecular basis of immunogenicity to botulinum neurotoxins and uses of the defined antigenic regions. *Toxicon* **2015**, *107 Pt A*, 50–58. [CrossRef]
126. Shone, C.; Agostini, H.; Clancy, J.; Gu, M.; Yang, H.H.; Chu, Y.; Johnson, V.; Taal, M.; McGlashan, J.; Brehm, J.; et al. Bivalent recombinant vaccine for botulinum neurotoxin types A and B based on a polypeptide comprising their effector and translocation domains that is protective against the predominant A and B subtypes. *Infect. Immun.* **2009**, *77*, 2795–2801. [CrossRef]
127. Binz, T.; Bade, S.; Rummel, A.; Kollewe, A.; Alves, J. Arg(362) and Tyr(365) of the botulinum neurotoxin type a light chain are involved in transition state stabilization. *Biochemistry* **2002**, *41*, 1717–1723. [CrossRef] [PubMed]
128. Agarwal, R.; Binz, T.; Swaminathan, S. Analysis of active site residues of botulinum neurotoxin E by mutational, functional, and structural studies: Glu335Gln is an apoenzyme. *Biochemistry* **2005**, *44*, 8291–8302. [CrossRef] [PubMed]
129. Ravichandran, E.; Janardhanan, P.; Patel, K.; Riding, S.; Cai, S.; Singh, B.R. In Vivo Toxicity and Immunological Characterization of Detoxified Recombinant Botulinum Neurotoxin Type A. *Pharm. Res.* **2016**, *33*, 639–652. [CrossRef]
130. Gu, S.; Rumpel, S.; Zhou, J.; Strotmeier, J.; Bigalke, H.; Perry, K.; Shoemaker, C.B.; Rummel, A.; Jin, R. Botulinum neurotoxin is shielded by NTNHA in an interlocked complex. *Science* **2012**, *335*, 977–981. [CrossRef]
131. Coronaviridae Study Group of the International Committee on Taxonomy of Viruses. The species Severe acute respiratory syndrome-related coronavirus: Classifying 2019-nCoV and naming it SARS-CoV-2. *Nat. Microbiol.* **2020**, *5*, 536–544. [CrossRef]
132. Brian, D.A.; Baric, R.S. Coronavirus genome structure and replication. *Curr. Top. Microbiol. Immunol.* **2005**, *287*, 1–30. [CrossRef]
133. Chilamakuri, R.; Agarwal, S. COVID-19: Characteristics and Therapeutics. *Cells* **2021**, *10*, 206. [CrossRef]
134. Wang, C.; Horby, P.W.; Hayden, F.G.; Gao, G.F. A novel coronavirus outbreak of global health concern. *Lancet* **2020**, *395*, 470–473. [CrossRef] [PubMed]
135. Hadj Hassine, I. COVID-19 vaccines and variants of concern: A review. *Rev. Med. Virol.* **2022**, *32*, e2313. [CrossRef]
136. Tregoning, J.S.; Flight, K.E.; Higham, S.L.; Wang, Z.; Pierce, B.F. Progress of the COVID-19 vaccine effort: Viruses, vaccines and variants versus efficacy, effectiveness and escape. *Nat. Rev. Immunol.* **2021**, *21*, 626–636. [CrossRef]
137. Sharif, N.; Alzahrani, K.J.; Ahmed, S.N.; Dey, S.K. Efficacy, Immunogenicity and Safety of COVID-19 Vaccines: A Systematic Review and Meta-Analysis. *Front. Immunol.* **2021**, *12*, 714170. [CrossRef]
138. Pollet, J.; Chen, W.H.; Strych, U. Recombinant protein vaccines, a proven approach against coronavirus pandemics. *Adv. Drug. Deliv. Rev.* **2021**, *170*, 71–82. [CrossRef] [PubMed]
139. Kumar, S.; Maurya, V.K.; Prasad, A.K.; Bhatt, M.L.B.; Saxena, S.K. Structural, glycosylation and antigenic variation between 2019 novel coronavirus (2019-nCoV) and SARS coronavirus (SARS-CoV). *Virusdisease* **2020**, *31*, 13–21. [CrossRef] [PubMed]
140. Drexler, J.F.; Corman, V.M.; Drosten, C. Ecology, evolution and classification of bat coronaviruses in the aftermath of SARS. *Antivir. Res.* **2014**, *101*, 45–56. [CrossRef] [PubMed]
141. Chen, Y.; Liu, Q.; Guo, D. Emerging coronaviruses: Genome structure, replication, and pathogenesis. *J. Med. Virol.* **2020**, *92*, 418–423. [CrossRef]

142. Lan, J.; Ge, J.; Yu, J.; Shan, S.; Zhou, H.; Fan, S.; Zhang, Q.; Shi, X.; Wang, Q.; Zhang, L.; et al. Structure of the SARS-CoV-2 spike receptor-binding domain bound to the ACE2 receptor. *Nature* **2020**, *581*, 215–220. [CrossRef] [PubMed]
143. Enose-Akahata, Y.; Wang, L.; Almsned, F.; Johnson, K.R.; Mina, Y.; Ohayon, J.; Wang, X.W.; Jacobson, S. The repertoire of CSF antiviral antibodies in patients with neuroinflammatory diseases. *Sci. Adv.* **2023**, *9*, eabq6978. [CrossRef]
144. Elko, E.A.; Nelson, G.A.; Mead, H.L.; Kelley, E.J.; Carvalho, S.T.; Sarbo, N.G.; Harms, C.E.; Le Verche, V.; Cardoso, A.A.; Ely, J.L.; et al. COVID-19 vaccination elicits an evolving, cross-reactive antibody response to epitopes conserved with endemic coronavirus spike proteins. *Cell Rep.* **2022**, *40*, 111022. [CrossRef]
145. Xiaojie, S.; Yu, L.; Lei, Y.; Guang, Y.; Min, Q. Neutralizing antibodies targeting SARS-CoV-2 spike protein. *Stem Cell Res.* **2020**, *50*, 102125. [CrossRef]
146. Novavax. Novavax's Nuvaxovid™ Receives Positive CHMP Opinion for Full Marketing Authorization for the Prevention of COVID in the EU. 2023. Available online: https://ir.novavax.com/press-releases/2023-05-26-Novavaxs-Nuvaxovid-TM-Receives-Positive-CHMP-Opinion-for-Full-Marketing-Authorization-for-the-Prevention-of-COVID-in-the-EU (accessed on 1 July 2023).
147. Heath, P.T.; Galiza, E.P.; Baxter, D.N.; Boffito, M.; Browne, D.; Burns, F.; Chadwick, D.R.; Clark, R.; Cosgrove, C.; Galloway, J.; et al. Safety and Efficacy of NVX-CoV2373 COVID-19 Vaccine. *N. Engl. J. Med.* **2021**, *385*, 1172–1183. [CrossRef]
148. Dunkle, L.M.; Kotloff, K.L.; Gay, C.L.; Anez, G.; Adelglass, J.M.; Barrat Hernandez, A.Q.; Harper, W.L.; Duncanson, D.M.; McArthur, M.A.; Florescu, D.F.; et al. Efficacy and Safety of NVX-CoV2373 in Adults in the United States and Mexico. *N. Engl. J. Med.* **2022**, *386*, 531–543. [CrossRef] [PubMed]
149. Toback, S.; Galiza, E.; Cosgrove, C.; Galloway, J.; Goodman, A.L.; Swift, P.A.; Rajaram, S.; Graves-Jones, A.; Edelman, J.; Burns, F.; et al. Safety, immunogenicity, and efficacy of a COVID-19 vaccine (NVX-CoV2373) co-administered with seasonal influenza vaccines: An exploratory substudy of a randomised, observer-blinded, placebo-controlled, phase 3 trial. *Lancet Respir. Med.* **2022**, *10*, 167–179. [CrossRef] [PubMed]
150. Trougakos, I.P.; Terpos, E.; Alexopoulos, H.; Politou, M.; Paraskevis, D.; Scorilas, A.; Kastritis, E.; Andreakos, E.; Dimopoulos, M.A. Adverse effects of COVID-19 mRNA vaccines: The spike hypothesis. *Trends Mol. Med.* **2022**, *28*, 542–554. [CrossRef] [PubMed]
151. Pardi, N.; Hogan, M.J.; Porter, F.W.; Weissman, D. mRNA vaccines—A new era in vaccinology. *Nat. Rev. Drug Discov.* **2018**, *17*, 261–279. [CrossRef]
152. Chen, J.; Chen, J.; Xu, Q. Current Developments and Challenges of mRNA Vaccines. *Annu. Rev. Biomed. Eng.* **2022**, *24*, 85–109. [CrossRef]
153. Rosa, S.S.; Prazeres, D.M.F.; Azevedo, A.M.; Marques, M.P.C. mRNA vaccines manufacturing: Challenges and bottlenecks. *Vaccine* **2021**, *39*, 2190–2200. [CrossRef]
154. Schlake, T.; Thess, A.; Fotin-Mleczek, M.; Kallen, K.J. Developing mRNA-vaccine technologies. *RNA Biol.* **2012**, *9*, 1319–1330. [CrossRef]
155. Teo, S.P. Review of COVID-19 mRNA Vaccines: BNT162b2 and mRNA-1273. *J. Pharm. Pract.* **2022**, *35*, 947–951. [CrossRef]
156. Wang, L.; Davis, P.B.; Kaelber, D.C.; Volkow, N.D.; Xu, R. Comparison of mRNA-1273 and BNT162b2 Vaccines on Breakthrough SARS-CoV-2 Infections, Hospitalizations, and Death during the Delta-Predominant Period. *JAMA* **2022**, *327*, 678–680. [CrossRef]
157. Garreffa, E.; Hamad, A.; O'Sullivan, C.C.; Hazim, A.Z.; York, J.; Puri, S.; Turnbull, A.; Robertson, J.F.; Goetz, M.P. Regional lymphadenopathy following COVID-19 vaccination: Literature review and considerations for patient management in breast cancer care. *Eur. J. Cancer* **2021**, *159*, 38–51. [CrossRef]
158. Shimabukuro, T. Allergic reactions including anaphylaxis after receipt of the first dose of Moderna COVID-19 vaccine—United States, 21 December 2020–10 January 2021. *Am. J. Transpl.* **2021**, *21*, 1326–1331. [CrossRef] [PubMed]
159. CDC COVID-19 Response Team; Food and Drug Administration. Allergic Reactions Including Anaphylaxis After Receipt of the First Dose of Moderna COVID-19 Vaccine—United States, 21 December 2020–10 January 2021. *MMWR Morb. Mortal. Wkly. Rep.* **2021**, *70*, 125–129. [CrossRef]
160. Buchan, S.A.; Alley, S.; Seo, C.Y.; Johnson, C.; Kwong, J.C.; Nasreen, S.; Thampi, N.; Lu, D.; Harris, T.M.; Calzavara, A.; et al. Myocarditis or Pericarditis Events After BNT162b2 Vaccination in Individuals Aged 12 to 17 Years in Ontario, Canada. *JAMA Pediatr.* **2023**, *177*, 410–418. [CrossRef] [PubMed]
161. Singer, M.E.; Taub, I.B.; Kaelber, D.C. Risk of Myocarditis from COVID-19 Infection in People Under Age 20: A Population-Based Analysis. *medRxiv* **2022**. medRxiv:2021.07.23.21260998. [CrossRef]
162. Ewer, K.; Sebastian, S.; Spencer, A.J.; Gilbert, S.; Hill, A.V.S.; Lambe, T. Chimpanzee adenoviral vectors as vaccines for outbreak pathogens. *Hum. Vaccin. Immunother.* **2017**, *13*, 3020–3032. [CrossRef]
163. Custers, J.; Kim, D.; Leyssen, M.; Gurwith, M.; Tomaka, F.; Robertson, J.; Heijnen, E.; Condit, R.; Shukarev, G.; Heerwegh, D.; et al. Vaccines based on replication incompetent Ad26 viral vectors: Standardized template with key considerations for a risk/benefit assessment. *Vaccine* **2021**, *39*, 3081–3101. [CrossRef]
164. Koger-Pease, C.; Perera, D.J.; Ndao, M. Recent Advances in the Development of Adenovirus-Vectored Vaccines for Parasitic Infections. *Pharmaceuticals* **2023**, *16*, 334. [CrossRef]
165. Francis, A.I.; Ghany, S.; Gilkes, T.; Umakanthan, S. Review of COVID-19 vaccine subtypes, efficacy and geographical distributions. *Postgrad. Med. J.* **2022**, *98*, 389–394. [CrossRef]

166. Food and Drug Administration (USFDA). Coronavirus (COVID-19) Update: FDA Limits Use of Janssen COVID-19 Vaccine to Certain Individuals. 2022. Available online: https://www.fda.gov/news-events/press-announcements/coronavirus-covid-19-update-fda-limits-use-janssen-covid-19-vaccine-certain-individuals (accessed on 1 July 2023).
167. Bozkurt, B. Shedding Light on Mechanisms of Myocarditis with COVID-19 mRNA Vaccines. *Circulation* **2023**, *147*, 877–880. [CrossRef]
168. Hajjo, R.; Sabbah, D.A.; Bardaweel, S.K.; Tropsha, A. Shedding the Light on Post-Vaccine Myocarditis and Pericarditis in COVID-19 and Non-COVID-19 Vaccine Recipients. *Vaccines* **2021**, *9*, 1186. [CrossRef]
169. Thran, M.; Mukherjee, J.; Ponisch, M.; Fiedler, K.; Thess, A.; Mui, B.L.; Hope, M.J.; Tam, Y.K.; Horscroft, N.; Heidenreich, R.; et al. mRNA mediates passive vaccination against infectious agents, toxins, and tumors. *EMBO Mol. Med.* **2017**, *9*, 1434–1447. [CrossRef] [PubMed]
170. Mukherjee, J.; Ondeck, C.A.; Tremblay, J.M.; Archer, J.; Debatis, M.; Foss, A.; Awata, J.; Erasmus, J.H.; McNutt, P.M.; Shoemaker, C.B. Intramuscular delivery of formulated RNA encoding six linked nanobodies is highly protective for exposures to three Botulinum neurotoxin serotypes. *Sci. Rep.* **2022**, *12*, 11664. [CrossRef]
171. Tomic, M.T.; Farr-Jones, S.; Syar, E.S.; Niemuth, N.; Kobs, D.; Hackett, M.J.; Espinoza, Y.; Martinez, Z.; Pham, K.; Snow, D.M.; et al. Neutralizing Concentrations of Anti-Botulinum Toxin Antibodies Positively Correlate with Mouse Neutralization Assay Results in a Guinea Pig Model. *Toxins* **2021**, *13*, 671. [CrossRef] [PubMed]
172. Liljeqvist, S.; Stahl, S. Production of recombinant subunit vaccines: Protein immunogens, live delivery systems and nucleic acid vaccines. *J. Biotechnol.* **1999**, *73*, 1–33. [CrossRef]
173. Mabrouk, M.T.; Huang, W.C.; Deng, B.; Li-Purcell, N.; Seffouh, A.; Ortega, J.; Ekin Atilla-Gokcumen, G.; Long, C.A.; Miura, K.; Lovell, J.F. Lyophilized, antigen-bound liposomes with reduced MPLA and enhanced thermostability. *Int. J. Pharm.* **2020**, *589*, 119843. [CrossRef]
174. Goldstein, N.; Bockstal, V.; Bart, S.; Luhn, K.; Robinson, C.; Gaddah, A.; Callendret, B.; Douoguih, M. Safety and Immunogenicity of Heterologous and Homologous 2-Dose Regimens of Adenovirus Serotype 26- and Modified Vaccinia Ankara-Vectored Ebola Vaccines: A Randomized, Controlled Phase 1 Study. *J. Infect. Dis.* **2022**, *226*, 595–607. [CrossRef] [PubMed]
175. Ura, T.; Okuda, K.; Shimada, M. Developments in Viral Vector-Based Vaccines. *Vaccines* **2014**, *2*, 624–641. [CrossRef]
176. Casimiro, D.R.; Bett, A.J.; Fu, T.M.; Davies, M.E.; Tang, A.; Wilson, K.A.; Chen, M.; Long, R.; McKelvey, T.; Chastain, M.; et al. Heterologous human immunodeficiency virus type 1 priming-boosting immunization strategies involving replication-defective adenovirus and poxvirus vaccine vectors. *J. Virol.* **2004**, *78*, 11434–11438. [CrossRef]
177. Xu, S.; Yang, K.; Li, R.; Zhang, L. mRNA Vaccine Era-Mechanisms, Drug Platform and Clinical Prospection. *Int. J. Mol. Sci.* **2020**, *21*, 6582. [CrossRef]
178. Jackson, N.A.C.; Kester, K.E.; Casimiro, D.; Gurunathan, S.; DeRosa, F. The promise of mRNA vaccines: A biotech and industrial perspective. *NPJ Vaccines* **2020**, *5*, 11. [CrossRef]
179. Hotez, P.J.; Bottazzi, M.E. Developing a low-cost and accessible COVID-19 vaccine for global health. *PLoS Neglected Trop. Dis.* **2020**, *14*, e0008548. [CrossRef]
180. Crommelin, D.J.A.; Anchordoquy, T.J.; Volkin, D.B.; Jiskoot, W.; Mastrobattista, E. Addressing the Cold Reality of mRNA Vaccine Stability. *J. Pharm. Sci.* **2021**, *110*, 997–1001. [CrossRef] [PubMed]
181. Muramatsu, H.; Lam, K.; Bajusz, C.; Laczko, D.; Kariko, K.; Schreiner, P.; Martin, A.; Lutwyche, P.; Heyes, J.; Pardi, N. Lyophilization provides long-term stability for a lipid nanoparticle-formulated, nucleoside-modified mRNA vaccine. *Mol. Ther.* **2022**, *30*, 1941–1951. [CrossRef] [PubMed]

*toxins*

MDPI

*Review*

# The Molecular Architecture and Mode of Action of *Clostridium perfringens* ε-Toxin

**Richard W. Titball**

Department of Biosciences, University of Exeter, Exeter EX4 4QD, UK; r.w.titball@exeter.ac.uk

**Abstract:** *Clostridium perfringens* ε-toxin has long been associated with a severe enterotoxaemia of livestock animals, and more recently, was proposed to play a role in the etiology of multiple sclerosis in humans. The remarkable potency of the toxin has intrigued researchers for many decades, who suggested that this indicated an enzymatic mode of action. Recently, there have been major breakthroughs by finding that it is a pore-forming toxin which shows exquisite specificity for cells bearing the myelin and lymphocyte protein (MAL) receptor. This review details the molecular structures of the toxin, the evidence which identifies MAL as the receptor and the possible roles of other cell membrane components in toxin binding. The information on structure and mode of action has allowed the functions of individual amino acids to be investigated and has led to the creation of mutants with reduced toxicity that could serve as vaccines. In spite of this progress, there are still a number of key questions around the mode of action of the toxin which need to be further investigated.

**Keywords:** *Clostridium perfringens*; ε-toxin; pore-forming toxin; structure; receptor

**Key Contribution:** This review highlights how an understanding of the structural basis of toxicity provides new insight into the prevention and treatments of diseases of animals and humans caused by *Clostridium perfringens* ε-toxin.

## 1. Introduction

*Clostridium perfringens* ε-toxin is produced by biotype B and D strains of the bacterium. These biotypes are associated with enteritis and enterotoxaemia in young livestock animals and especially in goats, calves and foals [1]. Enterotoxaemia in livestock is often fatal and death can occur soon after the first signs of disease [2]. The infection is typically acquired after the animal ingests spores of the bacterium, which then germinate in the gut. Disease is also associated with a dysbiosis in the gut, often as the result of a change in diet, which allows the multiplication of *C. perfringens* and the subsequent production of ε-toxin. The toxin crosses the gut wall, enters the bloodstream and then binds to target tissues. Histological analysis of ligated intestinal loops of sheep and goats exposed to ε-toxin showed necrosis of the colonic epithelium, and this may change the permeability of the intestine, allowing toxin translocation, possibly by opening tight junctions in the gut mucosa [3,4]. The main target organ is the brain, and entry of the toxin appears to be facilitated by an increase in the permeability of the blood–brain barrier which is associated with damage to vascular endothelial cells [2,5]. Within the brain, cells including neurons, astrocytes and oligodendrocytes are susceptible to ε-toxin [5]. Oligodendrocytes play a role in the myelination of axons, and some researchers have visualised changes to the lamellation of myelin in the brains of diseased animals [6]. Kidney damage is often reported, but it is unclear if this is a direct or an indirect consequence of disease caused by ε-toxin [7].

The ε-toxin gene (*etx*) is located on plasmids in both biotype B and D strains of *C. perfringens* [8] and the *etx* sequence, encoding a 311 amino acid protein, was first reported in 1992. There is no significant sequence homology between ε-toxin and sequences in the NCBI database [9]. The toxin is secreted as a relatively inactive 296 amino acid prototoxin [10], which is activated by proteolytic cleavage, for example by digestive proteases such as trypsin or chymotrypsin [11] or by *C. perfringens* λ-protease [12,13]. Proteolytic activation releases N- and C-terminal pro-peptides, with a consequent increase in toxicity of between 80- and 1000-fold [14,15]. The site of cleavage differs from protease to protease, and individual proteases may cleave at multiple sites. Therefore, the activated toxin may exist in a number of isoforms [14,15]. Whilst it was originally believed that the removal of both the N- and C-terminal pro-peptides was required for activation, a study which characterised recombinant forms of the toxin lacking the N- or C-terminal pro-peptides showed that activation is largely a consequence of the removal of the C-terminal pro-peptide [15]. It is not clear what the biological function of ε-toxin is in relation to the bacterium. However, like tetanus and botulinum toxin, it is possible that the potency reflects the benefit of rapidly killing the host and creating a local environment which is enriched for organic materials and then allows the proliferation of *C. perfringens*.

The mode of action of the toxin has, in the past, been a topic for speculation. In mice, the potency of the toxin, exceeded only by tetanus and botulinum toxins, was originally believed to indicate an enzymatic activity. The hope that gene cloning and sequencing would reveal a possible mode of action was dashed because ε-toxin did not show significant sequence homology to other toxins, or indeed to other proteins. Similarly, work to identify the receptor for ε-toxin led to diverse theories about the nature of the receptor molecule(s), and attempts to identity the receptor were, until recently, unsuccessful.

This manuscript reviews the major advances that have been made in our understanding of the mode of action and molecular basis of the toxicity of ε-toxin. These advances have provided new insight into this toxin, challenged the idea that enzymatic activity is necessary for high potency, renewed interest in the potential role of the toxin in multiple sclerosis (MS) in humans and led to new vaccine candidates for the prevention of disease caused by ε-toxin.

## 2. *C. perfringens* ε-Toxin Is a Member of the Aerolysin Family of Pore-Forming Toxins

The molecular structure of the monomeric form of ε-toxin was first reported in 2004 [9] and revealed an elongated (100 Å × 20 Å × 20 Å) molecule with three domains, composed mainly from β-sheet (Figure 1a). Although there was no significant sequence similarity with aerolysin, a pore-forming toxin produced by *Aeromonas hydrophila*, the overall structures of ε-toxin [9] and aerolysin [16] showed similarities. However, ε-toxin comprises three domains, while there are four in aerolysin (Figure 1). This finding suggests that ε-toxin is a member of the aerolysin family of β-pore-forming proteins. This family includes proteins from diverse origins (Table 1) which show similarity of structure and mode of action, although there is little amino acid sequence homology between them. Proteins belonging to the aerolysin family are typically secreted as water-soluble monomers which bind to receptors on target cells and oligomerise to form ion-selective pores in the target cell membrane. Much of our understanding of structure–function relationships in members of this group are based on the pioneering studies with aerolysin, the prototypic member of this group. Bacterial toxins in the aerolysin family, such as ε-toxin, aerolysin, monalysin and the binary larvicide AB toxin (BinA/BinB) play a role in the etiology of microbial diseases in hosts ranging from insects to fish and mammals. Eukaryotic proteins in the aerolysin family, such as lysenin and *Laetiporus sulphureus* lectin, may be host defence proteins.

**Table 1.** Comparison of epsilon toxin with selected membrane active proteins belonging to the aerolysin group of pore-forming proteins.

| Sub-Class | ETX/MTX-2 | Aerolysin | Non-Etx | Lysenin | Monalysin | Toxin_10 |
|---|---|---|---|---|---|---|
| Example | Epsilon-toxin | Aerolysin | Laetiporus sulphureus lectin | Lysenin | Monalysin | BinA/BinB (Tpp1/Tpp2) |
| Origin | *C. perfringens* (Gram-positive bacterium) | *A. hydrophila* (Gram-negative bacterium) | *Laetiporus Sulphureus* (edible mushroom) | *Eisenia fetida* (earthworm) | *Pseudomonas entomophila* (Gram-negative bacterium) | *Lysinibacillus sphaericus* (Gram-positive bacterium) |
| Disease associations | Enterotoxaemia in animals. (MS in humans) | Motile Aeromonas Septicaemia in fish. Enteric disease in humans | Reported anticancer properties; used in traditional medicine | Believed to be a host defence protein | Lethal infection of insects | Lethal to some mosquito larvae species. Potential anti-cancer agent |
| Organisation |  |  |  |  |  | 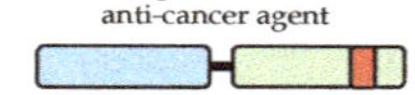 |
| Pfam | PF03318 | PF01117 | PF03318 | Not assigned | Not assigned | PF15431 |
| Monomers forming pore | 7 | 7 | Not known | 9 | Not known | Not known |
| Receptor | Myelin and lymphocyte protein (MAL) | GPI-anchored proteins | N-acetyllactosamine | Sphingomyelin | Not reported | Cpm1/Cqm1 amylomaltase (GPI anchored protein) |
| Activation | N- and C-terminal peptide removal | C-terminal peptide removal | Not reported | Not reported | N-terminal peptide removal | Insect gut proteases |
| Haemolytic activity | Haemolytic only for human red cells | Haemolytic towards a range of mammalian red cells. | Haemolytic towards a range of mammalian red cells | Haemolytic towards a range of mammalian red cells | Haemolytic to human red cells | Not reported |
| Toxicity in animals | $LD_{50}$ ~50 ng/kg in mice | $LD_{50}$ ~10 µg/kg in mice | Not lethal to zebrafish embryos at 600 µg/mL | Coelomic fluid containing lysenin is toxic to some vertebrates | Lethal to *Drosophila* sp. | Lethal to *Culex* sp. |

Graphics (not to scale) show the linear organisation of domains in the proteins. Receptor-binding domain is shown in cyan and orange. Pore-forming module (pale green and pink) containing the amphipathic loop (red). In the case of epsilon toxin and aerolysin, the receptor-binding domain is composed from two non-contiguous regions.

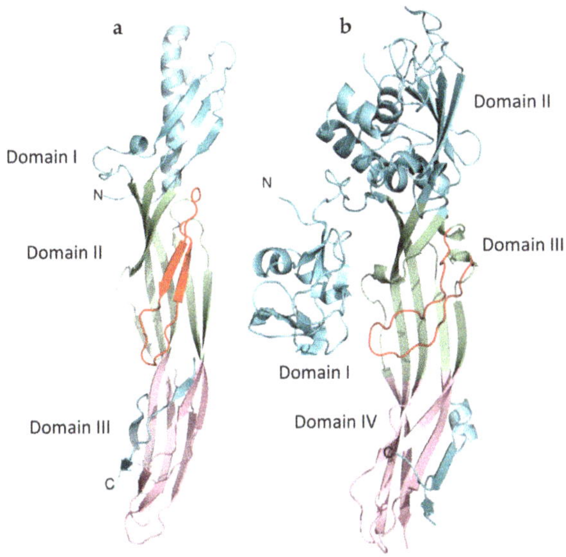


**Figure 1.** Comparison of the molecular structures of ε-toxin (**a**) and aerolysin (**b**). N- and C-termini of the proteins are marked. The domains are coloured according to function, with receptor-binding domains shaded cyan and the pore-forming modules composed from domains shaded green and pink. The amphipathic loop which forms the inner part of the transmembrane β-barrel is shaded red and located in domain II in ε-toxin and domain III in aerolysin. The C-terminal pro-peptide, removed by proteolytic cleavage is also shaded cyan. Figure modified from [17].

Although proteins in the aerolysin family show some similarities, there are important differences between the toxins in this group. This has resulted in the proposal for a number of groupings (Table 1), based largely on the differences in the presence and organisation of individual domains and assignment in the pfam database of protein families [18]. *C. perfringens* ε-toxin can be assigned to the ETX/MTX-2 group in the aerolysin family, alongside *Bacillus thuringiensis* Cry51 (Mpp51Aa1), *Bacillus thuringiensis* Cry23 (Mpp23Aa1) and *Bacillus thuringiensis* parasporin-2 proteins [18]. Cry51 and Cry23 are insecticidal toxins, with activity towards coleopterans. Parasporin-2 has attracted attention because of its activity towards cancer cells [19]. The alpha-toxin from *Clostridium septicum* is also a member of the aerolysin group of pore-forming toxins but in the absence of structural information, and a low level of amino acid sequence similarity with ε-toxin, it is unclear whether it is also a member of the ETX/MTX-2 group [20].

Except for monalysin, proteins in the aerolysin family possess a distinct receptor-binding domain (RBD) (Table 1). The possession of a RBD, which is encoded non-contiguously, is a key feature of the ETX/MTX-2 group, and this distinguishes this group from aerolysin which has two RBDs (domains I and II). In the non-Etx, lysenin and Toxin_10 sub-classes, the single RBD is encoded contiguously. In ε-toxin and most other aerolysin family proteins, the RBD is located at towards the N-terminus of the protein, but in lysenin, it is located at the C-terminus. The receptor-binding domains (RBDs) appear to be responsible for the differences in target cell specificity and toxicity, which are reported with different members of the aerolysin family.

All of the proteins possess a pore-forming module (PFM), which can include one or two domains. The PFM includes β-strands and a loop, which is often termed the amphipathic loop and inserts into the target cell membrane to form the inner part of the pore [21]. The amphipathic loop is another conserved feature of aerolysin family proteins. Like ε-toxin, many of the proteins in the aerolysin family are produced as relatively inactive precursors, activated by the proteolytic removal of peptides at the N- or C-terminal. All are presumed to form pores in target cells, and this explains their reported haemolytic activity.

However, the number of sub-units forming the pore (where known) ranges from seven to nine monomers. The pore itself is formed from a β-barrel which spans the membrane with a collar at the upper end and a "rivet" which forms on the inside of the membrane and these lock the pore into place [22]. The remainder of this article considers the structural biology of ε-toxin in more detail, making comparisons where appropriate with data available for other toxins in the aerolysin family.

## 3. The Receptor-Binding Domain

Human red cells, but not red cells from sheep, goat, mouse, rat, cow, dog, monkey, rabbit or horse are susceptible to lysis by ε-toxin [23]. In contrast, aerolysin causes the lysis of erythrocytes from a broad range of species, with the rat cells susceptible to sub-nanomolar concentrations [24]. Most cell lines tested are resistant to ε-toxin [25]. Some are highly sensitive to the toxin, for example Madin Darby Canine Kidney (MDCK) cells [26], with a $CT_{50}$ dose of 1–10 nM [27,28] and MOLT-4 human T cells with a $CT_{50}$ dose of 11 nM [28]. Aerolysin is active towards a broad range of cell lines, and when tested using a sensitive cell line, aerolysin is a more potent cytolytic agent than ε-toxin with EL4 T-cells killed by concentrations as low as 10 pM [24]. However, aerolysin and ε-toxin differ markedly in their toxicity to small mammals. The lethal dose of aerolysin in mice is approximately 10 μg/kg [29], whereas ε-toxin is at least 100 times more potent [30]. This indicates that the lethality of ε-toxin, compared to that of aerolysin, is not simply a consequence of increased cytotoxicity. Rather, it reflects different target cell specificities.

Whilst the RBDs of ε-toxin (domain I) and aerolysin (domain II) show some general structural similarities, they also show a number of differences. Compared to ε-toxin domain I, aerolysin domain II possesses additional helices, extended loops and two additional strands, resulting in a poor root-mean-square deviation of atomic positions (RMSD) when comparing the structures (1.9 Å over 23 of 62 C-alphas). In aerolysin, domain II is critical for binding to glycan residues on glycosylphosphatidylinositol (GPI)-anchored protein receptors [31] and more specifically, it binds to mannose residues in the glycan core which is tagged onto the lipid moiety of the GPI anchor [32]. Experimental data show that ε-toxin does not bind to GPI-anchored proteins, and although O-linked oligosaccharides have previously been suggested to play a role in the binding of ε-toxin to cells, the protein acceptor for these sugars was not identified in this study [33].

In ε-toxin, a cluster of aromatic amino acids on the upper face of domain I (Figure 2) appear to be involved in binding to MDCK cells, including Y16, Y20, Y29, Y30, Y36, Y196 and F199 [34–37]. However, a F199E variant protein showed profound changes in thermal stability compared to wild-type ε-toxin, suggesting significant changes in folding and conformation which may account for the reduction in binding [35,36]. A W199H mutant did not show a significant reduction in toxicity [37]. Therefore, it is unclear if F199 plays a direct role in binding. Amino acids Y16 and Y20 play a minor role in cell recognition [34,36]. The remaining amino acids, Y29, Y30, Y36 and Y196, are located on loops at the top of domain I of ε-toxin, bordering a cleft. This is similar in location and organisation to the cleft in domain II of aerolysin, which has been shown to be the site of mannose-6-phosphate binding in the glycan core on CD52, a GPI-anchored protein [32]. A recent modelling study by Kumar et al. [38] has suggested additional aromatic amino acids of ε-toxin toxin that may play a role in binding, and this now requires testing experimentally. The side chains of aromatic amino acids are often involved in binding to carbohydrate-binding proteins [39]. It is possible that the side chains of Y29, Y30, Y36 and Y196 of ε-toxin stack against the pyranose rings of sugars. However, other members of the aerolysin family use tyrosines to bind to the amino acid side chains of protein [40] or sphingomyelin receptors [41].

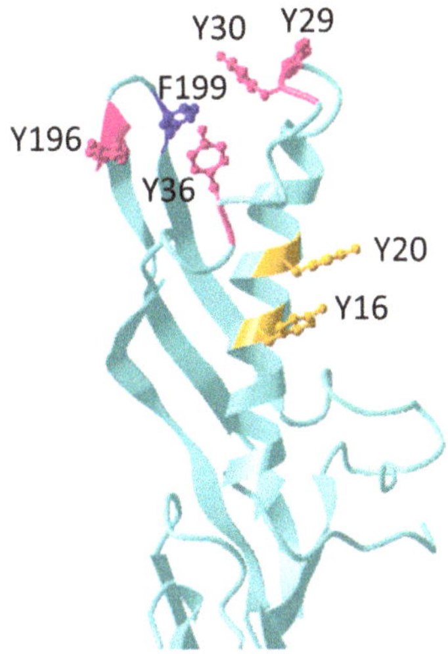


**Figure 2.** Locations of amino acids in domain I of ε-toxin that are believed to play major (pink) or minor (yellow) roles in binding to receptors. The location of F199 is also shown in blue, though an F199H variant retained toxicity, and an F199E variant appeared to show conformational changes, making it difficult to assess the role of this amino acid in binding.

Whilst domain I clearly plays a key role in receptor binding by ε-toxin, a possible β-octyl glucoside binding site has also been identified in domain III [34]. In the pore form of the toxin, the residues involved in β-octyl glucoside binding [42] are some distance from the outer leaflet of the target cell membrane. The properties of variants with substituted amino acids involved in hydrogen bonding to the glucose unit (E61, V72 and T93) of β-octyl glucoside have not been reported. This makes it difficult to understand the biological significance of this site.

## 4. The Pore-Forming Module

The molecular structures of the monomeric [9] and pore forms [42] of ε-toxin indicate that it is a β-pore-forming toxin, and this is supported by previous biophysical studies which demonstrate the formation of heptameric complexes of monomers in membranes exposed to ε-toxin [15]. Though the pores formed by ε-toxin are sometimes reported to be anion selective [43,44], the molecular structure of the pore shows that there is no overall charge in the lumen of the pore [42]. This is consistent with the finding that both cations and anions can access the pore, with changes in intracellular $K^+$, $Cl^-$, $Ca^{2+}$ and $Na^+$ levels in cells exposed to ε-toxin [44–46]. It is believed that these changes in ion concentrations in cells trigger the depletion of cellular ATP, stimulate an AMP-activated protein kinase and cause changes to the permeability of the mitochondrial membrane with translocation of the apoptosis-inducing factor [46].

Domains II and III of ε-toxin constitute the pore-forming module, and these domains show structural similarity to domains III and IV of aerolysin. In the ε-toxin monomer, the amphipathic loop that forms the lower part of the transmembrane β-barrel [9] is sandwiched between anti-parallel β-sheets in domain II. However, β-strands flanking the amphipathic loop (residues 95–173) also contribute to the pore [42]. Like aerolysin, it is clear that there are significant differences in the molecular structures of the monomeric [9] and pore [42] forms of ε-toxin (Figures 1a and 3). For aerolysin, the structural data on receptor binding and from mutants trapped at the pre-pore stages have allowed a model to be developed describing changes in the structure during the pore formation process [32]. This prototypic model is valuable for understanding the possible stages involved in pore formation by ε-toxin. An outer β-barrel (residues 63–94, 174–198 and 235–260) forms the so-called collar, surrounding the upper part of the inner β-barrel. Together, the outer β-barrel and the concentric inner β-barrel form the double β-barrel fold, which is characteristic of aerolysin family proteins [21,22,47] (Figure 3).

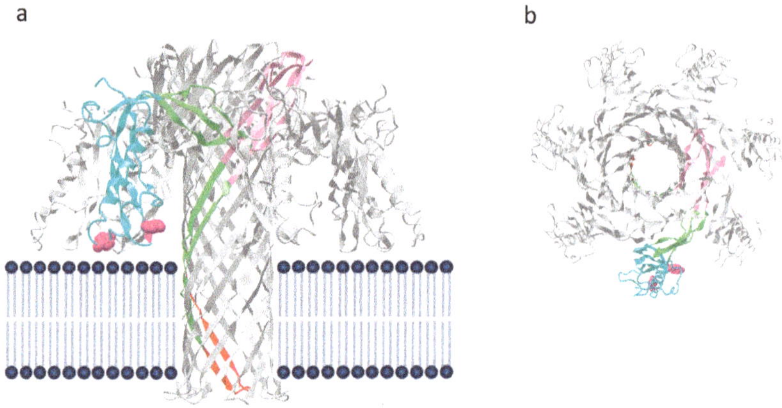


**Figure 3.** The molecular structure of the ε-toxin pore. A single protomer is shown coloured with domain I cyan, domain 2 pink and domain 3 green. The amphipathic loop in domain 2 is shaded red. The side chains of the receptor-binding residues Y30 and Y196 are show in dark pink. Panel (**a**) The pore is shown in the predicted position in the membrane bilayer (dark blue). (**b**) top view of the pore.

The ε-toxin pore shows overall structural similarities with the lysenin [48,49] and aerolysin [50] pores, and at its widest point measures approximately 24 Å. At the narrowest point, the diameter is only 12 Å, because of several lysine side chains that occupy the lumen. On the inside of the target cell membrane, T132, V133, P134 and F135 form a β-turn which enables a rivet to form, anchoring the barrel of the pore on the inside of the cell membrane.

The mechanism of pore formation and assembly seems likely to be similar to that of aerolysin. Like aerolysin, ε-toxin forms pores inefficiently in solution, but pores are formed readily on, and in, the target cell membrane. The receptors for both aerolysin and ε-toxin are more abundant in lipid rafts in the cell membrane, and this enables the monomers to bind, become locally concentrated [51] and assemble into structured pre-pores on the cell surface. The structure of the aerolysin pre-pore at different stages has been reported, and this shows an inverted mushroom structure with the stalk extending outwards from the membrane which becomes progressively flattened as the pre-pore moves towards pore formation [32]. The heptamer is believed to insert spontaneously into the lipid bilayer, possibly at the edge of lipid rafts [51]. For ε-toxin, the RBD is believed to remain attached to the receptor during the pre-pore and pore formation processes, and the structure of the RBD changes little, though there is movement of domain I relative to domain II [42]. For aerolysin, the most profound conformational changes in the PFM likely occur as it unfolds and inserts into the bilayer [21], and this also seems likely for ε-toxin.

## 5. Receptors

Whilst aerolysin appears to bind to a wide range of cells, and to a wide range of GPI-anchored proteins in the target cell membrane [51], a number of early studies demonstrated that ε-toxin binds selectively to host cells in vivo. For example, using peroxidase-labelled antibody to ε-toxin Buxton [52] showed that the pro-toxin accumulated in the brain, kidney and liver with some binding to the luminal surface of vascular endothelial cells.

Lipid rafts in cell membranes bring together monomers of the toxin to form the pore [53,54] and some studies implicate lipids such as sulfatides or phosphatidyl serine in binding ε-toxin [54]. This is consistent with the enrichment of these lipids in lipid rafts. Sulfatides are abundant in the myelin sheath, kidney tissues, the gastrointestinal tract and T-cells [55]. Phosphatidylserine is a major phospholipid in the cerebral cortex, though phosphatidyl serine is normally associated with the inner leaflet of the membrane. Cholesterol, which plays a structural role in lipid rafts, is also necessary for the binding of ε-toxin to detergent-resistant membranes [53]. All of this is consistent with ε-toxin binding to lipid rafts. It is possible that ε-toxin can insert into lipid monolayers or bilayers, if the concentration of toxin is sufficiently high to drive multimer formation in the absence of a high-affinity receptor. This would explain why some researchers report that liposomes or monolayers show changes in permeability when exposed to the toxin [46,56,57].

The identity of a protein receptor has been investigated by several groups over the past decades. Unlike aerolysin, the receptor for ε-toxin is not a GPI-anchored protein [45]. Some workers suggested that the receptor is a sialoglyoprotein, since synaptosome membranes treated with pronase or neuraminidase showed reduced binding of the toxin [58].

Two candidate protein receptors have been identified to date (Table 2). The Hepatitis A Virus Receptor (HAVCR1, also known at TIM-1) protein is the receptor for several human pathogenic viruses, and Ivie et al. [59] showed that reducing the expression of HAVCR1 in either MDCK or ACHN cells increased their resistance to ε-toxin. The authors were also able to demonstrate binding of ε-toxin to the extracellular domain of HAVCR1 in vitro [59]. Tyrosine residues in domain I which have been shown to play a role in cell binding (Y29, Y30, Y36 and Y196) appear to play a role in binding to the extracellular domain of HAVCR1 [36]. However, although the expression of HAVCR1 in a toxin-resistant cell line resulted in the expression of the protein on the cell surface, the cells did not bind or become susceptible to ε-toxin. The finding that HAVCR1 expression did not confer toxin sensitivity in cells might indicate that this protein forms part of a receptor complex, but is not the main toxin receptor.

**Table 2.** Comparison of HAVCR1 and MAL as receptors for epsilon toxin.

| Property | Protein Receptor | |
|---|---|---|
| | **HAVCR1** | **MAL** |
| Tissue distribution corelates with susceptibility to toxin? | No. HAVCR1 is expressed in all organs. Some resistant cells do not express HAVCR1. Some sensitive cells express HAVCR1 | Yes, except normal endothelial cells do not express MAL |
| Membrane localization | NR | In lipid rafts |
| Expression in cells confers toxin-sensitive phenotype? | No | Yes |
| Reduced expression increases resistance to toxin? | Yes | Yes |
| Gene inactivation confers whole animal resistance to toxin? | NR | Yes, in mice |
| Physical interaction of toxin with protein demonstrated? | Yes | Yes |
| ε-toxin amino acids involved in receptor binding | Y29, Y30, Y36, Y196 | Y30, Y196 |

NR = not reported.

An important advance was made by Rumah et al. in 2015 [60] who showed that CHO cells, which are normally resistant to ε-toxin, become highly sensitive when engineered to express MAL. Moreover, MAL-knockout ($Mal^{-/-}$) mice which no longer express MAL in any cells are highly resistant to intoxication. Also, unlike oligodendrocytes from wild-type mice, oligodendrocytes cultured from $Mal^{-/-}$ mice fail to bind to ε-toxin and are toxin resistant [61]. Subsequently, other researchers have confirmed that the expression of MAL by cultured cells is associated with susceptibility to the toxin [27,28,56,62–64]. MAL is associated with lipid rafts in cells, and MAL is reported to oligomerise under some conditions [65]. Evidence of a physical interaction between ε-toxin and MAL was provided by Blanch et al. [28] who showed that the complex could be immunoprecipitated from cells that had previously been exposed to toxin. Collectively, these findings provide a substantial body of evidence that MAL is a major receptor for ε-toxin, but do not preclude an additional role for HAVCR1, lipids or glycans associated with the proteins in binding.

MAL is a 17kDa protein expressed in polarised epithelial cells and in myelin forming cells [66]. MAL is associated with the plasma membrane and also with membranes within cells, and has an essential role in vesicular transport and plasma membrane targeting in a range of cell types [66]. The tissue distribution of MAL does broadly map to the pathology seen in intoxicated animals; it is expressed strongly in epithelial cells in the kidney and in oligodendrocytes which play a role in myelin deposition. It is also expressed in T-cells (especially CD4+ T-cells) and red cells in humans, but not in these cell types in other mammals [66]. In addition to the full-length protein, a number of isoforms exist, and according to a preprint, in humans, an 11kDa isoform is present on red cells which is believed to lack the exon-III encoded region [67].

MAL from different animal species appears to bind to ε-toxin with different affinities. This is well illustrated by comparing the behaviour of CHO cells expressing different MAL; in two independent studies, cells expressing human MAL were at least 10 times less sensitive to ε-toxin than cells expressing rat or sheep MAL [27,60]. More distantly related MAL, for example from zebra fish, appears not to bind to ε-toxin at all, and zebra fish engineered to express human MAL become toxin-sensitive [64]. It also appears that individual amino acids in ε-toxin play different roles in binding to MAL from different species [27,56,60]. The finding that MAL is found on T-cells and red cells in humans but not in other animals indicates that these cells would likely bind to ε-toxin in the blood stream, and this would then limit the amount of toxin that was available to cross the blood–brain barrier and bind to cells in the brain. This might explain why acute toxicity associated with extensive damage to the brain is not seen in humans [68].

The detailed molecular interactions between MAL and ε-toxin are not clear, and there is no structure available of MAL, or the MAL-ε-toxin complex. A study in which the potential interactions between ε-toxin and MAL were modelled [38] might provide information on the interaction, and other studies have provided experimental data. MAL is expressed in a number of isoforms in cells, and the 11kDa isoform of MAL on human red cells [67] would include a binding motif. Rumah et al. [60] have suggested that ε-toxin binds to the predicted extracellular loop 3 of MAL identified by Rubio-Ramos et al. [66], because an insertion into this predicted loop abolished the binding of ε-toxin. However, according to Rubio-Ramos et al. [66], this predicted loop is identical in human and rat MAL, suggesting that it is not responsible for the differences in the susceptibilities of cells expressing human or rat MAL.

Overall, although MAL appears to be an important receptor for ε-toxin, it seems likely that there are other binding partners for ε-toxin, including HAVCR1 and lipids. Possibly, MAL is involved in the transport of HAVCR1 to the cell membrane, and might form a complex with MAL at the cell surface. The observation that ε-toxin binds to endothelial cells lining blood vessels [52], yet MAL is not known to be expressed in normal endothelial cells [66], requires further investigation. Glycans, which may be attached to MAL or HAVCR1, might also play a key role in binding. It is conceivable that different binding partners exist, with molecules such as lipids binding with low affinity, and MAL binding with high affinity to ε-toxin. The determination of the fine molecular structure of the ε-toxin and receptor complex is now an important goal in order to address these questions.

## 6. Exploiting Information on the Mode of Action

In the past two decades, and because of its potency, ε-toxin has received attention as a potential biowarfare or bioterrorism agent. However, a number of findings suggest that the extensive pathology seen in intoxicated animals is less likely to be seen in humans exposed to ε-toxin. There are two reports of the isolation of *C. perfringens* able to produce ε-toxin from the gut of individuals suffering from enteric disease [69]. Neither individual was reported to show signs of intoxication, and there was no report of death as a consequence of the infection [69]. Also, as highlighted above, the pathogenesis of disease in humans caused by ε-toxin would likely be different to that seen in animals because of differences in MAL distribution and affinity for the toxin.

However, it is possible that epsilon toxin causes chronic disease in humans, and a paper published by Murrell in 1986 [70] first outlined the hypothesis that there was a link between exposure to ε-toxin and the later development of MS in humans. This hypothesis was largely based around the observation that five of eight researchers who had worked on swayback disease (an ovine neurological disease) developed MS [71], while the probability of this chance happening was 1 in $10^9$. Murrell considered that these researchers could have been exposed to *C. perfringens* strains which produced ε-toxin. In a broader epidemiological study, Murrell confirmed an association between exposure to sheep or sheep products and the development of MS [70]. However, he could not find antibodies to ε-toxin in individuals who developed MS. More recently, the finding that the receptor for ε-toxin is MAL, which is located in the myelin sheath of neurons, triggered renewed interest in this hypothesis. Subsequently, antibodies which react with ε-toxin in MS patients have been reported [72,73]. Using the PCR, two separate studies [74,75] have shown that the *etx* gene is found at higher frequencies in faecal samples from MS patients compared to matched controls (61% vs. 13% and 14% vs. 3% in the two studies). Isolates cultured from MS patients have been shown to produce functional ε-toxin [72,75]. Moreover, when present, the abundance of *C. perfringens* strains which possess the *etx* gene was higher in MS patients than in controls [75]. There was no pathology reported in control subjects who tested positive for *etx*. It is possible that individuals become colonised after ingesting foodstuffs, such as unpasteurised milk, that contains *etx*-positive *C. perfringens* [76]. Using the experimental autoimmune encephalomyelitis model of MS researchers showed that ε-toxin might play a role in triggering the disease by opening the blood–brain barrier [75]. Overall, the possibility that ε-toxin plays a role in MS warrants further research.

Information on the structure of ε-toxin is being exploited in several ways. Variant proteins with reduced abilities to bind to cells, located in domain I, are one route to the development of low-toxicity variants that could be exploited as vaccines [27,37,77–80]. The mutation of other residues, for example F118 located in the transmembrane loop [81], also results in a reduction in toxicity. Also, the targeted mutation of selected residues to prolines [82] can reduce toxicity, but this may be because the substitution disrupts the structure of the toxin rather than because the amino acid being targeted plays a key role in toxicity. The mutation of H149 and F92 residues within or surrounding the β-octyl glucoside binding site results in a modest reduction in toxicity [34,83]. Other residues such as Y71 [83,84], which is located in domain III, have also been targeted for mutagenesis and may play a role in pore formation.

An alternative approach to devising a vaccine against pore-forming toxins has been proposed by Hu et al. [85] working with *Staphylococcus aureus* α-hemolysin. This involved allowing the toxin to bind to nanoparticles coated with host cell membrane to generate "nanotoxoids". It is unclear if the bound toxin formed pores, but the resultant nanoparticles did not show toxicity or reversion to toxicity. Similar findings have been reported with *C. perfringens* ε-toxin [86]. The nanoparticles coated with *S. aureus* α-hemolysin were immunogenic, eliciting neutralising antibody, but it is not clear if this was antibody directed against the monomer or the pore forms of the toxin, and whether antibody was able to neutralise toxicity after the toxin had bound to target cell membranes. The finding that the structures of the monomer and pore forms of the toxin are markedly different suggests the limited conservation of B-cell epitopes and highlights the need to understand more fully how neutralising antibodies can protect against pore-forming toxins.

A longer-term goal might be to engineer the properties of the pore for applications in biotechnology or medicine. Knapp et al. [46] have reported that targeted amino acid substitutions to the amino acid lining the pore can change the selectivity of the pore towards cations. Interestingly, the ion-selectivity of the pore correlated poorly with cytotoxicity.

## 7. Conclusions

The past decade has seen major advances in our understanding of the molecular basis of the toxicity of ε-toxin. *C. perfringens* ε-toxin is one of the most potent bacterial toxins known, exceeded only by botulinum and tetanus toxin. For many years, this high toxicity was thought to indicate an enzymatic mode of action. We now know that ε-toxin does not have an enzymatic mode of action, and its potency is a consequence of targeted damage to cells bearing the myelin and lymphocyte protein (MAL) receptor. It is also clear that, like other pore-forming toxins, remarkable changes in the structure of the protein take place as it becomes bound to and then inserts into target cell membranes. However, we do not know the detail of the intermediate steps in the pore-forming process.

This finding might open opportunities to engineer the toxin with different target cell specificities, to selectively kill some cells or to enhance drug delivery into cells. We are now close to understanding how the toxin interacts with membranes, but it is not clear if MAL alone can serve as a receptor, or whether other membrane components are required, and if so, what role they play in binding. Understanding the interaction between ε-toxin and MAL is a priority to address this question. The finding that MAL appears to be the major receptor protein has renewed the interest in the possible role of ε-toxin in MS, and additional work is required to investigate this possibility. This is a significant challenge because animals other than humans are not known to develop MS, and the distribution and affinity of receptors is different in humans compared to other mammals.

**Funding:** This research received no external funding.

**Institutional Review Board Statement:** Not applicable.

**Informed Consent Statement:** Not applicable.

**Data Availability Statement:** Data reported in this review can be found in the relevant cited article, available at https://pubmed.ncbi.nlm.nih.gov/ (accessed on 12 January 2024).

**Conflicts of Interest:** The author is a shareholder in One Health Vaccines, which has an interest in vaccines against *Clostridium perfringens* epsilon toxin.

## References

1. Songer, J.G. Clostridial enteric diseases of domestic animals. *Clin. Microbiol. Rev.* **1996**, *9*, 216–234. [CrossRef] [PubMed]
2. Finnie, J.W. Pathogenesis of brain damage produced in sheep by *Clostridium perfringens* type D epsilon toxin: A review. *Aust. Vet. J.* **2003**, *81*, 219–221. [CrossRef] [PubMed]
3. Goldstein, J.; Morris, W.E.; Loidl, C.F.; Tironi-Farinati, C.; McClane, B.A.; Uzal, F.A.; Fernandez Miyakawa, M.E. *Clostridium perfringens* epsilon toxin increases the small intestinal permeability in mice and rats. *PLoS ONE* **2009**, *4*, e7065. [CrossRef] [PubMed]
4. Fernandez Miyakawa, M.E.; Uzal, F.A. The early effects of *Clostridium perfringens* type D epsilon toxin in ligated intestinal loops of goats and sheep. *Vet. Res. Commun.* **2003**, *27*, 231–241. [CrossRef] [PubMed]
5. Freedman, J.C.; McClane, B.A.; Uzal, F.A. New insights into *Clostridium perfringens* epsilon toxin activation and action on the brain during enterotoxemia. *Anaerobe* **2016**, *41*, 27–31. [CrossRef] [PubMed]
6. Finnie, J.W. Ultrastructural changes in the brain of mice given *Clostridium perfringens* type D epsilon toxin. *J. Comp. Pathol.* **1984**, *94*, 445–452. [CrossRef]
7. Giannitti, F.; García, J.P.; Adams, V.; Armendano, J.I.; Beingesser, J.; Rood, J.I.; Uzal, F.A. Experimental acute *Clostridium perfringens* type D enterotoxemia in sheep is not characterized by specific renal lesions. *Vet. Pathol.* **2023**, *60*, 412–419. [CrossRef]
8. Canard, B.; Saint-Joanis, B.; Cole, S.T. Genomic diversity and organization of virulence genes in the pathogenic anaerobe *Clostridium perfringens*. *Mol. Microbiol.* **1992**, *6*, 1421–1429. [CrossRef]
9. Cole, A.R.; Gibert, M.; Popoff, M.; Moss, D.S.; Titball, R.W.; Basak, A.K. *Clostridium perfringens* epsilon-toxin shows structural similarity to the pore-forming toxin aerolysin. *Nat. Struct. Mol. Biol.* **2004**, *11*, 797–798. [CrossRef]
10. McDonel, J.L. Toxins of *Clostridium perfringens* types A, B, C, D and E. In *Pharmacology of Bacterial Toxins*; Dorner, F., Drews, J., Eds.; Pergamon Press: Oxford, UK, 1986; pp. 477–517.
11. Bhown, A.S.; Habeerb, A.F. Structural studies on epsilon-prototoxin of *Clostridium perfringens* type D. Localization of the site of tryptic scission necessary for activation to epsilon-toxin. *Biochem. Biophys. Res. Commun.* **1977**, *78*, 889–896. [CrossRef]
12. Jin, F.; Matsushita, O.; Katayama, S.; Jin, S.; Matsushita, C.; Minami, J.; Okabe, A. Purification, characterization, and primary structure of *Clostridium perfringens* lambda-toxin, a thermolysin-like metalloprotease. *Infect. Immun.* **1996**, *64*, 230–237. [CrossRef]
13. Minami, J.; Katayama, S.; Matsushita, O.; Matsushita, C.; Okabe, A. Lambda-toxin of *Clostridium perfringens* activates the precursor of epsilon-toxin by releasing its N- and C-terminal peptides. *Microbiol. Immunol.* **1997**, *41*, 527–535. [CrossRef] [PubMed]

14. Worthington, R.W.; Mulders, M.S. Physical changes in the epsilon prototoxin molecule of *Clostridium perfringens* during enzymatic activation. *Infect. Immun.* **1977**, *18*, 549–551. [CrossRef]
15. Miyata, S.; Matsushita, O.; Minami, J.; Katayama, S.; Shimamoto, S.; Okabe, A. Cleavage of a C-terminal peptide is essential for heptamerization of *Clostridium perfringens* epsilon-toxin in the synaptosomal membrane. *J. Biol. Chem.* **2001**, *276*, 13778–13783. [CrossRef]
16. Parker, M.W.; Buckley, J.T.; Postma, J.P.M.; Tucker, A.D.; Leonard, K.; Pattus, F.; Tsernoglou, D. Structure of the *Aeromonas toxin* proaerolysin in its water-soluble and membrane-channel states. *Nature* **1994**, *367*, 292–295. [CrossRef]
17. Bokori-Brown, M.; Savva, C.G.; Fernandes da Costa, S.P.; Naylor, C.E.; Basak, A.K.; Titball, R.W. Molecular basis of toxicity of *Clostridium perfringens* epsilon toxin. *FEBS J.* **2011**, *278*, 4589–4601. [CrossRef] [PubMed]
18. Lacomel, C.J.; Dunstone, M.A.; Spicer, B.A. Branching out the aerolysin, ETX/MTX-2 and Toxin_10 family of pore forming proteins. *J. Invertebr. Pathol.* **2021**, *186*, 107570. [CrossRef] [PubMed]
19. Akiba, T.; Okumura, S. Parasporins 1 and 2: Their structure and activity. *J. Invertebr. Pathol.* **2017**, *142*, 44–49. [CrossRef]
20. Ballard, J.; Crabtree, J.; Roe, B.A.; Tweten, R.K. The primary structure of *Clostridium septicum* alpha-toxin exhibits similarity with that of *Aeromonas hydrophila* aerolysin. *Infect. Immun.* **1995**, *63*, 340–344. [CrossRef]
21. Podobnik, M.; Kisovec, M.; Anderluh, G. Molecular mechanism of pore formation by aerolysin-like proteins. *Philos. Trans. R. Soc. Lond. B Biol. Sci.* **2017**, *372*, 20160209. [CrossRef]
22. Cirauqui, N.; Abriata, L.A.; van der Goot, F.G.; Dal Peraro, M. Structural, physicochemical and dynamic features conserved within the aerolysin pore-forming toxin family. *Sci. Rep.* **2017**, *7*, 13932. [CrossRef] [PubMed]
23. Gao, J.; Xin, W.; Huang, J.; Ji, B.; Gao, S.; Chen, L.; Kang, L.; Yang, H.; Shen, X.; Zhao, B.; et al. Hemolysis in human erythrocytes by *Clostridium perfringens* epsilon toxin requires activation of P2 receptors. *Virulence* **2018**, *9*, 1601–1614. [CrossRef] [PubMed]
24. Diep, D.B.; Nelson, K.L.; Lawrence, T.S.; Sellman, B.R.; Tweten, R.K.; Buckley, J.T. Expression and properties of an aerolysin—*Clostridium septicum* alpha toxin hybrid protein. *Mol. Microbiol.* **1999**, *31*, 785–794. [CrossRef] [PubMed]
25. Shortt, S.J.; Titball, R.W.; Lindsay, C.D. An assessment of the in vitro toxicology of *Clostridium perfringens* type D epsilon-toxin in human and animal cells. *Hum. Exp. Toxicol.* **2000**, *19*, 108–116. [CrossRef] [PubMed]
26. Knight, P.A.; Queminet, J.; Blanchard, J.H.; Tilleray, J.H. In vitro tests for the measurement of clostridial toxins, toxoids and antisera. II. Titration of *Clostridium perfringens* toxins and antitoxins in cell culture. *Biologicals* **1990**, *18*, 263–270. [CrossRef] [PubMed]
27. Morcrette, H.; Bokori-Brown, M.; Ong, S.; Bennett, L.; Wren, B.W.; Lewis, N.; Titball, R.W. *Clostridium perfringens* epsilon toxin vaccine candidate lacking toxicity to cells expressing myelin and lymphocyte protein. *NPJ Vaccines* **2019**, *4*, 32. [CrossRef] [PubMed]
28. Blanch, M.; Dorca-Arévalo, J.; Not, A.; Cases, M.; Gómez de Aranda, I.; Martínez-Yélamos, A.; Martínez-Yélamos, S.; Solsona, C.; Blasi, J. The cytotoxicity of epsilon toxin from *Clostridium perfringens* on lymphocytes is mediated by MAL protein expression. *Mol. Cell. Biol.* **2018**, *38*, e00086-18. [CrossRef] [PubMed]
29. Buckley, J.T.; Halasa, L.N.; Lund, K.D.; MacIntyre, S. Purification and some properties of the hemolytic toxin aerolysin. *Can. J. Biochem.* **1981**, *59*, 430–435. [CrossRef] [PubMed]
30. Gill, D.M. Bacterial toxins: A table of lethal amounts. *Microbiol. Rev.* **1982**, *46*, 86–94. [CrossRef]
31. Hong, Y.J.; Ohishi, K.; Inoue, N.; Kang, J.Y.; Shime, H.; Horiguchi, Y.; van der Goot, F.G.; Sugimoto, N.; Kinoshita, T. Requirement of N-glycan on GPI-anchored proteins for efficient binding of aerolysin but not *Clostridium septicum* alpha-toxin. *EMBO J.* **2002**, *21*, 5047–5056. [CrossRef]
32. Degiacomi, M.T.; Iacovache, I.; Pernot, L.; Chami, M.; Kudryashev, M.; Stahlberg, H.; van der Goot, F.G.; Dal Peraro, M. Molecular assembly of the aerolysin pore reveals a swirling membrane-insertion mechanism. *Nat. Chem. Biol.* **2013**, *9*, 623–629. [CrossRef] [PubMed]
33. Dorca-Arévalo, J.; Martín-Satué, M.; Blasi, J. Characterization of the high affinity binding of epsilon toxin from *Clostridium perfringens* to the renal system. *Vet. Microbiol.* **2012**, *157*, 179–189. [CrossRef] [PubMed]
34. Bokori-Brown, M.; Kokkinidou, M.C.; Savva, C.G.; Fernandes da Costa, S.; Naylor, C.E.; Cole, A.R.; Moss, D.S.; Basak, A.K.; Titball, R.W. *Clostridium perfringens* epsilon toxin H149A mutant as a platform for receptor binding studies. *Protein Sci.* **2013**, *22*, 650–659. [CrossRef] [PubMed]
35. Kang, J.; Gao, J.; Yao, W.; Kang, L.; Gao, S.; Yang, H.; Ji, B.; Li, P.; Liu, J.; Yao, J.; et al. F199E substitution reduced toxicity of *Clostridium perfringens* epsilon toxin by depriving the receptor binding capability. *Hum. Vaccin. Immunother.* **2017**, *13*, 1598–1608. [CrossRef] [PubMed]
36. Ivie, S.E.; McClain, M.S. Identification of amino acids important for binding of *Clostridium perfringens* epsilon toxin to host cells and to HAVCR1. *Biochemistry* **2012**, *51*, 7588–7595. [CrossRef] [PubMed]
37. Li, Q.; Xin, W.; Gao, S.; Kang, L.; Wang, J. A low-toxic site-directed mutant of *Clostridium perfringens* ε-toxin as a potential candidate vaccine against enterotoxemia. *Hum. Vaccines Immunother.* **2013**, *9*, 2386–2392. [CrossRef] [PubMed]
38. Kumar, S.; Behera, S.K.; Gururaj, K.; Chaurasia, A.; Murmu, S.; Prabha, R.; Angadi, U.B.; Pawaiya, R.S.; Rai, A. In silico mutation of aromatic with aliphatic amino acid residues in *Clostridium perfringens* epsilon toxin (ETX) reduces its binding efficiency to caprine myelin and lymphocyte (MAL) protein receptors. *J. Biomol. Struct. Dyn.* **2024**, *42*, 2257–2269. [CrossRef] [PubMed]
39. Quiocho, F.A. Carbohydrate-binding proteins: Tertiary structures and protein-sugar interactions. *Annu. Rev. Biochem.* **1986**, *55*, 287–315. [CrossRef]

40. Yelland, T.S.; Naylor, C.E.; Bagoban, T.; Savva, C.G.; Moss, D.S.; McClane, B.A.; Blasig, I.E.; Popoff, M.; Basak, A.K. Structure of a *C. perfringens* enterotoxin mutant in complex with a modified Claudin-2 extracellular loop 2. *J. Mol. Biol.* **2014**, *426*, 3134–3147. [CrossRef]
41. De Colibus, L.; Sonnen, A.F.; Morris, K.J.; Siebert, C.A.; Abrusci, P.; Plitzko, J.; Hodnik, V.; Leippe, M.; Volpi, E.; Anderluh, G.; et al. Structures of lysenin reveal a shared evolutionary origin for pore-forming proteins and its mode of sphingomyelin recognition. *Structure* **2012**, *20*, 1498–1507. [CrossRef]
42. Savva, C.G.; Clark, A.R.; Naylor, C.E.; Popoff, M.R.; Moss, D.S.; Basak, A.K.; Titball, R.W.; Bokori-Brown, M. The pore structure of *Clostridium perfringens* epsilon toxin. *Nat. Commun.* **2019**, *10*, 2641. [CrossRef] [PubMed]
43. Nestorovich, E.M.; Karginov, V.A.; Bezrukov, S.M. Polymer partitioning and ion selectivity suggest asymmetrical shape for the membrane pore formed by epsilon toxin. *Biophys. J.* **2010**, *99*, 782–789. [CrossRef] [PubMed]
44. Petit, L.; Maier, E.; Gibert, M.; Popoff, M.R.; Benz, R. *Clostridium perfringens* epsilon toxin induces a rapid change of cell membrane permeability to ions and forms channels in artificial lipid bilayers. *J. Biol. Chem.* **2001**, *276*, 15736–15740. [CrossRef]
45. Chassin, C.; Bens, M.; de Barry, J.; Courjaret, R.; Bossu, J.L.; Cluzeaud, F.; Ben Mkaddem, S.; Gibert, M.; Poulain, B.; Popoff, M.R.; et al. Pore-forming epsilon toxin causes membrane permeabilization and rapid ATP depletion-mediated cell death in renal collecting duct cells. *Am. J. Physiol. Renal Physiol.* **2007**, *293*, F927–F937. [CrossRef] [PubMed]
46. Knapp, O.; Maier, E.; Piselli, C.; Benz, R.; Hoxha, C.; Popoff, M.R. Central residues of the amphipathic β-hairpin loop control the properties of *Clostridium perfringens* epsilon-toxin channel. *Biochim. Biophys. Acta Biomembr.* **2020**, *1862*, 183364. [CrossRef] [PubMed]
47. Johnstone, B.A.; Christie, M.P.; Morton, C.J.; Parker, M.W. X-ray crystallography shines a light on pore-forming toxins. *Methods Enzymol.* **2021**, *649*, 1–46. [CrossRef] [PubMed]
48. Bokori-Brown, M.; Martin, T.G.; Naylor, C.E.; Basak, A.K.; Titball, R.W.; Savva, C.G. Cryo-EM structure of lysenin pore elucidates membrane insertion by an aerolysin family protein. *Nat. Commun.* **2016**, *7*, 11293. [CrossRef] [PubMed]
49. Podobnik, M.; Savory, P.; Rojko, N.; Kisovec, M.; Wood, N.; Hambley, R.; Pugh, J.; Wallace, E.J.; McNeill, L.; Bruce, M.; et al. Crystal structure of an invertebrate cytolysin pore reveals unique properties and mechanism of assembly. *Nat. Commun.* **2016**, *7*, 11598. [CrossRef] [PubMed]
50. Iacovache, I.; De Carlo, S.; Cirauqui, N.; Dal Peraro, M.; van der Goot, F.G.; Zuber, B. Cryo-EM structure of aerolysin variants reveals a novel protein fold and the pore-formation process. *Nat. Commun.* **2016**, *7*, 12062. [CrossRef]
51. Abrami, L.; Fivaz, M.; van der Goot, F.G. Adventures of a pore-forming toxin at the target cell surface. *Trends Microbiol.* **2000**, *8*, 168–172. [CrossRef]
52. Buxton, D. The use of an immunoperoxidase technique to investigate by light and electron microscopy the sites of binding of *Clostridium welchii* type-D epsilon toxin in mice. *J. Med. Microbiol.* **1978**, *11*, 289–292. [CrossRef]
53. Miyata, S.; Minami, J.; Tamai, E.; Matsushita, O.; Shimamoto, S.; Okabe, A. *Clostridium perfringens* epsilon-toxin forms a heptameric pore within the detergent-insoluble microdomains of Madin-Darby canine kidney cells and rat synaptosomes. *J. Biol. Chem.* **2002**, *277*, 39463–39468. [CrossRef] [PubMed]
54. Gil, C.; Dorca-Arévalo, J.; Blasi, J. *Clostridium perfringens* epsilon toxin binds to membrane lipids and its cytotoxic action depends on sulfatide. *PLoS ONE* **2015**, *10*, e0140321. [CrossRef] [PubMed]
55. Takahashi, T.; Suzuki, T. Role of sulfatide in normal and pathological cells and tissues. *J. Lipid Res.* **2012**, *53*, 1437–1450. [CrossRef]
56. Marshall, S.; McGill, B.; Morcrette, H.; Winlove, C.P.; Chimerel, C.; Petrov, P.G.; Bokori-Brown, M. Interaction of *Clostridium perfringens* epsilon toxin with the plasma membrane: The role of amino acids Y42, Y43 and H162. *Toxins* **2022**, *14*, 757. [CrossRef] [PubMed]
57. Nagahama, M.; Hara, H.; Fernandez-Miyakawa, M.; Itohayashi, Y.; Sakurai, J. Oligomerization of *Clostridium perfringens* epsilon-toxin is dependent upon membrane fluidity in liposomes. *Biochemistry* **2006**, *45*, 296–302. [CrossRef] [PubMed]
58. Nagahama, M.; Sakurai, J. High-affinity binding of *Clostridium perfringens* epsilon-toxin to rat brain. *Infect. Immun.* **1992**, *60*, 1237–1240. [CrossRef] [PubMed]
59. Ivie, S.E.; Fennessey, C.M.; Sheng, J.; Rubin, D.H.; McClain, M.S. Gene-trap mutagenesis identifies mammalian genes contributing to intoxication by *Clostridium perfringens* epsilon-toxin. *PLoS ONE* **2011**, *6*, e17787. [CrossRef] [PubMed]
60. Rumah, K.R.; Ma, Y.; Linden, J.R.; Oo, M.L.; Anrather, J.; Schaeren-Wiemers, N.; Alonso, M.A.; Fischetti, V.A.; McClain, M.S.; Vartanian, T. The myelin and lymphocyte protein MAL Is Required for binding and activity of *Clostridium perfringens* epsilon-toxin. *PLoS Pathog.* **2015**, *11*, e1004896. [CrossRef]
61. Linden, J.R.; Ma, Y.; Zhao, B.; Harris, J.M.; Rumah, K.R.; Schaeren-Wiemers, N.; Vartanian, T. *Clostridium perfringens* epsilon toxin causes selective death of mature oligodendrocytes and central nervous system demyelination. *mBio* **2015**, *16*, e02513. [CrossRef]
62. Geng, Z.; Huang, J.; Kang, L.; Gao, S.; Yuan, Y.; Li, Y.; Wang, J.; Xin, W.; Wang, J. *Clostridium perfringens* epsilon toxin binds to erythrocyte MAL receptors and triggers phosphatidylserine exposure. *J. Cell Mol. Med.* **2020**, *24*, 7341–7352. [CrossRef]
63. Linden, J.R.; Flores, C.; Schmidt, E.F.; Uzal, F.A.; Michel, A.O.; Valenzuela, M.; Dobrow, S.; Vartanian, T. *Clostridium perfringens* epsilon toxin induces blood brain barrier permeability via caveolae-dependent transcytosis and requires expression of MAL. *PLoS Pathog.* **2019**, *15*, e1008014. [CrossRef] [PubMed]
64. Adler, D.; Linden, J.R.; Shetty, S.V.; Ma, Y.; Bokori-Brown, M.; Titball, R.W.; Vartanian, T. *Clostridium perfringens* epsilon toxin compromises the blood-brain barrier in a humanized zebrafish model. *iScience* **2019**, *15*, 39–54. [CrossRef]

65. Ramnarayanan, S.P.; Tuma, P.L. MAL, but not MAL2, expression promotes the formation of cholesterol-dependent membrane domains that recruit apical proteins. *Biochem. J.* **2011**, *439*, 497–504. [CrossRef] [PubMed]
66. Rubio-Ramos, A.; Labat-de-Hoz, L.; Correas, I.; Alonso, M.A. The MAL protein, an integral component of specialized membranes, in normal cells and cancer. *Cells* **2021**, *10*, 1065. [CrossRef]
67. Rumah, K.R.; Eleso, O.E.; Fischetti, V.A. Human blood exposure to *Clostridium perfringens* epsilon toxin may shed light on erythrocyte fragility during active multiple sclerosis. *bioRxiv* **2019**, 789123. [CrossRef]
68. Titball, R.W.; Lewis, N.; Nicholas, R. Is *Clostridium perfringens* epsilon toxin associated with multiple sclerosis? *Mult. Scler.* **2023**, *29*, 1057–1063. [CrossRef]
69. Popoff, M.R. Epsilon toxin: A fascinating pore-forming toxin. *FEBS J.* **2011**, *278*, 4602–4615. [CrossRef] [PubMed]
70. Murrell, T.G.; O'Donoghue, P.J.; Ellis, T. A review of the sheep-multiple sclerosis connection. *Med. Hypotheses* **1986**, *19*, 27–39. [CrossRef]
71. Dean, G.; McDougall, E.I.; Elian, M. Multiple sclerosis in research workers studying swayback in lambs: An updated report. *J. Neurol. Neurosurg. Psychiatry* **1985**, *48*, 859–865. [CrossRef]
72. Rumah, K.R.; Linden, J.; Fischetti, V.A.; Vartanian, T. Isolation of *Clostridium perfringens* type B in an individual at first clinical presentation of multiple sclerosis provides clues for environmental triggers of the disease. *PLoS ONE* **2013**, *8*, e76359. [CrossRef] [PubMed]
73. Wagley, S.; Bokori-Brown, M.; Morcrette, H.; Malaspina, A.; D'Arcy, C.; Gnanapavan, S.; Lewis, N.; Popoff, M.R.; Raciborska, D.; Nicholas, R.; et al. Evidence of *Clostridium perfringens* epsilon toxin associated with multiple sclerosis. *Mult. Scler.* **2018**, *25*, 653–660. [CrossRef] [PubMed]
74. Maryam, S.A.; Babak, K.; Nima, B. Determination of type and molecular identity of *Clostridium perfringens* isolated from patients with multiple sclerosis (MS). *J. Adv. Biomed. Sci.* **2022**, *12*, 325–335.
75. Ma, Y.; Sannino, D.; Linden, J.R.; Haigh, S.; Zhao, B.; Grigg, J.B.; Zumbo, P.; Dündar, F.; Butler, D.J.; Profaci, C.P.; et al. Epsilon toxin-producing *Clostridium perfringens* colonize the MS gut and epsilon toxin overcomes immune privilege. *J. Clin. Investig.* **2023**, *133*, e163239. [CrossRef] [PubMed]
76. Gupta, T.B.; Brightwell, G. Identification and characterisation of spore-forming bacteria in bovine raw milk collected from four dairy farms in New Zealand. *Dairy* **2023**, *4*, 650–671. [CrossRef]
77. Bokori-Brown, M.; Hall, C.A.; Vance, C.; Fernandes da Costa, S.P.; Savva, C.G.; Naylor, C.E.; Cole, A.R.; Basak, A.K.; Moss, D.S.; Titball, R.W. *Clostridium perfringens* epsilon toxin mutant Y30A-Y196A as a recombinant vaccine candidate against enterotoxemia. *Vaccine* **2014**, *32*, 2682–2687. [CrossRef] [PubMed]
78. Du, J.; Wang, T.; Xu, L.; Wang, C.; Liu, Y.; Pan, C.; Chen, X.; Zhu, Z.; Luo, Y.; Yin, C. *Clostridium perfringens* epsilon prototoxin mutant rpETX(Y30A/Y71A/H106P/Y196A) as a vaccine candidate against enterotoxemia. *Vaccine* **2023**, *41*, 4762–4770. [CrossRef]
79. Peng, X.; Li, X.; Peng, G.; Feng, L.; Jiang, Y.; Luo, Y. Recombinant unpurified rETX(H106P)/CTB-rETX(Y196E) protects rabbits against *Clostridium perfringens* epsilon toxin. *J. Vet. Med. Sci.* **2021**, *83*, 441–446. [CrossRef]
80. Yao, W.; Kang, J.; Kang, L.; Gao, S.; Yang, H.; Ji, B.; Li, P.; Liu, J.; Xin, W.; Wang, J. Immunization with a novel *Clostridium perfringens* epsilon toxin mutant rETX(Y196E)-C confers strong protection in mice. *Sci. Rep.* **2016**, *6*, 24162. [CrossRef]
81. Dorca-Arevalo, J.; Pauillac, S.; Diaz-Hidalgo, L.; Martin-Satue, M.; Popoff, M.R.; Blasi, J. Correlation between in vitro cytotoxicity and in vivo lethal activity in mice of epsilon toxin mutants from *Clostridium perfringens*. *PLoS ONE* **2014**, *9*, e102417. [CrossRef]
82. Oyston, P.C.F.; Payne, D.W.; Havard, H.L.; Williamson, E.D.; Titball, R.W. Production of a non-toxic site-directed mutant of *Clostridium perfringens* epsilon-toxin which induces protective immunity in mice. *Microbiology* **1998**, *144*, 333–341. [CrossRef] [PubMed]
83. Garg, L.; Gopal, K.; Dixit, A. Recombinant Vaccine against Clostridium Perfringens Infection and Epsilon Toxin Intoxication. WIPO Patent WO2012004645A1, 12 January 2012.
84. Jiang, Z.; Chang, J.; Wang, F.; Qi, Y.; Li, Y.; Yu, D.; Yu, L. Etx-Y71A as a non-toxic mutant of *Clostridium perfringens* epsilon toxin induces protective immunity in mice and sheep. *Vaccine* **2020**, *38*, 6553–6561. [CrossRef] [PubMed]
85. Hu, C.M.; Fang, R.H.; Luk, B.T.; Zhang, L. Nanoparticle-detained toxins for safe and effective vaccination. *Nat. Nanotechnol.* **2013**, *8*, 933–938. [CrossRef] [PubMed]
86. Pudineh Moarref, M.; Alimolaei, M.; Emami, T.; Koohi, M.K. Development and evaluation of cell membrane-based biomimetic nanoparticles loaded by *Clostridium perfringens* epsilon toxin: A novel vaccine delivery platform for Clostridial-associated diseases. *Nanotoxicology* **2023**, *17*, 420–431. [CrossRef]

*toxins*

MDPI

*Article*

# Proteome and Phosphoproteome Profiling Reveal the Toxic Mechanism of *Clostridium perfringens* Epsilon Toxin in MDCK Cells

**Nan Yue [1,†], Jing Huang [2,†], Mingxin Dong [3,†], Jiaxin Li [1], Shan Gao [1], Jing Wang [1], Yingshuang Wang [4], Dongxue Li [1], Xi Luo [1], Tingting Liu [1], Songyang Han [1], Lina Dong [1], Ming Chen [1], Jinglin Wang [1], Na Xu [5,*], Lin Kang [1] and Wenwen Xin [1,*]**

1 State Key Laboratory of Pathogen and Biosecurity, Institute of Microbiology and Epidemiology, AMMS, Beijing 100071, China; yn18795116327@163.com (N.Y.); ljx1658675586@163.com (J.L.); gaoshan845@163.com (S.G.); amms_wj@163.com (J.W.); dongxueli@163.com (D.L.); luoxi18907093302@163.com (X.L.); liutingtingwushi@163.com (T.L.); hansongyang0.0@outlook.com (S.H.); donglina2024@163.com (L.D.); chenming3901@163.com (M.C.); wjlwjl0801@sina.com (J.W.); kanglin@bmi.ac.cn (L.K.)

2 National Key Laboratory of Intelligent Tracking and Forecasting for Infectious Diseases, National Institute of Environmental Health, Chinese Center for Disease Control and Prevention, Beijing 100020, China; huangjing@nieh.chinacdc.cn

3 Changchun Veterinary Research Institute, Chinese Academy of Agricultural Science, Changchun 130122, China; realdmx311@163.com

4 Health Bureau of Nanguan District, Changchun 130022, China; wangys19@mails.jlu.edu.cn

5 Academic Affairs Office, Jilin Medical University, Jilin 132013, China

* Correspondence: xuna@jlmu.edu.cn (N.X.); xinwenwen@bmi.ac.cn (W.X.); Tel.: +86-10-66948643 (W.X.)

† These authors contributed equally to this work.

**Abstract:** Epsilon toxin (ETX), a potential agent of biological and toxic warfare, causes the death of many ruminants and threatens human health. It is crucial to understand the toxic mechanism of such a highly lethal and rapid course toxin. In this study, we detected the effects of ETX on the proteome and phosphoproteome of MDCK cells after 10 min and 30 min. A total of 44 differentially expressed proteins (DEPs) and 588 differentially phosphorylated proteins (DPPs) were screened in the 10 min group, while 73 DEPs and 489 DPPs were screened in the 30 min group. ETX-induced proteins and phosphorylated proteins were mainly located in the nucleus, cytoplasm, and mitochondria, and their enrichment pathways were related to transcription and translation, virus infection, and intercellular junction. Meanwhile, the protein–protein interaction network screened out several hub proteins, including SRSF1/2/6/7/11, SF3B1/2, NOP14/56, ANLN, GTPBP4, THOC2, and RRP1B. Almost all of these proteins were present in the spliceosome pathway, indicating that the spliceosome pathway is involved in ETX-induced cell death. Next, we used RNAi lentiviruses and inhibitors of several key proteins to verify whether these proteins play a critical role. The results confirmed that SRSF1, SF3B2, and THOC2 were the key proteins involved in the cytotoxic effect of ETX. In addition, we found that the common upstream kinase of these key proteins was SRPK1, and a reduction in the level of SRPK1 could also reduce ETX-induced cell death. This result was consistent with the phosphorylated proteomics analysis. In summary, our study demonstrated that ETX induces phosphorylation of SRSF1, SF3B2, THOC2, and SRPK1 proteins on the spliceosome pathway, which inhibits normal splicing of mRNA and leads to cell death.

**Keywords:** epsilon toxin; proteomics; phospho-proteomics; toxic mechanism; SRSF1; spliceosome

**Key Contribution:** ETX is produced by *Clostridium perfringens* types B and D. Studies have shown that *Botulinum* and *Clostridium tetanus* neurotoxins are recognized as the most lethal toxins, and ETX is second only to them. However, how ETX induces illness and death remains unclear. Previous studies have shown that the spliceosome and its pathway play key roles in the development and progression of various cancers and that phosphorylation of SRPK1 is required for SARS-CoV-2 replication. However, we found a relationship between the spliceosome pathway and toxin pathogenicity that, to our knowledge, has never been reported before. These findings also provide data support for the search for early markers of ETX exposure and targets for treatment.

## 1. Introduction

Epsilon toxin (ε-toxin, ETX), produced by *Clostridium perfringens* types B and D [1], is the causative agent of enterotoxaemia in animals [2,3] and is also considered to be related to multiple sclerosis in humans, causing human health threats and economic losses to the global livestock industry [4,5]. Due to its high toxicity, ETX has been classified as a potential bioterrorism or bio-warfare agent by the Centers for Disease Control and Prevention of the United States. ETX has been listed as one of the potential biological warfare agents by the Centers for Disease Control and Prevention of the United States and is also listed as a category B bioterrorism threat [6]. In in vivo challenge experiments in mice, it was found that the median lethal dose of ETX to mice was 70–110 ng/kg [7,8].

ETX is composed of 311 amino acids and has a molecular weight of 32.7 kDa. At first, it is simply an inactive proETX that can be activated by extracellular serine-type enzymes [9]. Specifically, these enzymes hydrolyze to remove N-terminal and C-terminal peptides, allowing ETX to be converted to an activated form with toxic effects. Studies have shown that ETX can cause Madin Darby canine kidney (MDCK) cells to swell and rupture, eventually leading to cell death [10]. In recent years, an increasing number of cells have been found to be susceptible to ETX including the Caucasian renal leiomyoblastoma (G-402) cell line [11], the human kidney cell line ACHN [12], the murine renal cortical collecting duct principal cell line mpkCCDcl4 [13], and human erythrocytes [14]. However, MDCK cells are still the cells most used in ETX studies because they are the most sensitive [15].

The exact mechanisms of action of ETX on target cells are barely understood. Previous studies have shown that ETX can directly form pores in the target cell membrane and is a typical pore-forming toxin [10,16–18]. The formation of pores causes the exchange of materials between the inside and outside of the cell to be disrupted, which leads to the death of the cell [12]. However, some recent studies have found that ETX may also exert toxic effects through other pathways of action [12,14,19,20]. For example, Wioland has reported that ETX can cause oligodendrocyte demyelination [20]. Gao demonstrated that ETX can cause hemolysis in human erythrocytes by activating P2 receptors on the surface of erythrocytes [14]. Interestingly, P2 × 7 and P2Y13 inhibitors inhibited hemolysis but not ETX space formation [14]. In addition, it has been suggested that oxidation of nucleotide-sensitive ICln chloride channels contributes to ETX-induced hemolysis [21]. These studies together indicate that ETX leads to cell death mode not only through the formation of membrane pores but also through a variety of other effective pathways.

Most of the previous studies on ETX were on the pore-forming mechanism of ETX itself and on various mutants. To our knowledge, this is the first joint proteomic and phosphorylation proteomic study of the effects of ETX on MDCK cells. In this study, we profiled cell responses to ETX intoxication by performing quantitative proteomics and phospho-proteomics using MDCK cells to find clues about the pathogenesis of ETX.

## 2. Results

### 2.1. Pre-Assessment of Phosphorylation on MDCK Cells and Workflow

To determine the point times of ETX effects on MDCK cells, immunoblotting was performed. The result is shown in Figure 1A; after being exposed to ETX for 10 min, the phosphorylated proteins were reduced. However, at 30 min, the number of phosphorylated proteins was increased. Therefore, we chose the 10 min and 30 min intervals for the following experiments. The flow chart of quantitative proteome and phosphorylated proteome is shown in Figure 1B.

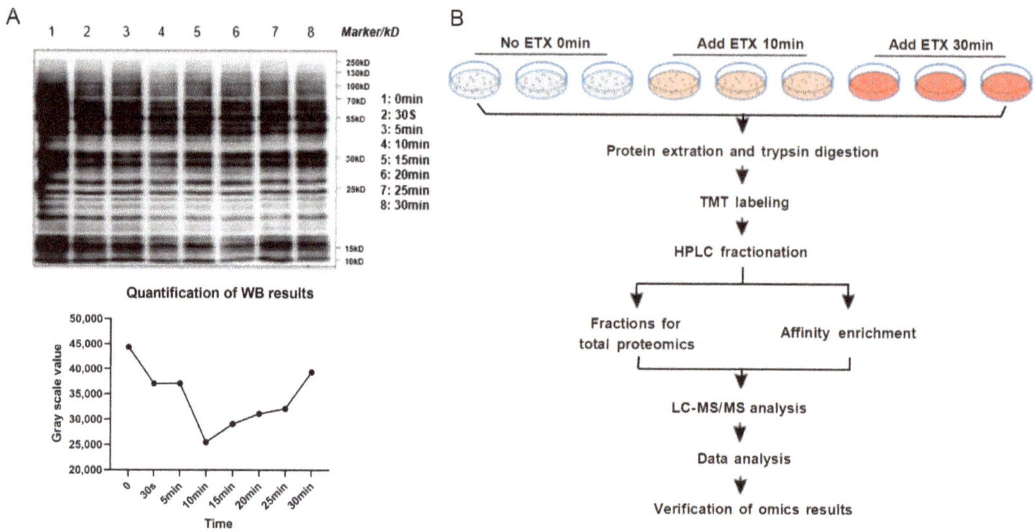


**Figure 1.** Pre-assessment of phosphorylation and workflow. (**A**) Western blotting result with pan anti-phosphotyrosine; WB results were quantified using image J software (version number Java 1.8.0–112). Three times independent gel was run for phosphoproteins. (**B**) Workflow of global proteome and phosphorylome analyses of ETX-treated MDCK cells. Three independent LC-MS/MS biology experiments were performed for each group.

### 2.2. *Screening of Differentially Expressed Proteins and Differentially Phosphorylated Proteins*

By TMT (Tandem Mass Tags) labeling, affinity enrichment, and mass spectrometry, 6086 proteins were identified, of which 5397 proteins contained quantitative information (Table S1). We have identified 9989 phosphorylation sites, respectively, in 3199 of the proteins, of which 4916 loci with quantitative information correspond to 1688 proteins (Table S2). This indicates that the larger the protein molecular weight, the smaller the coverage (Figure 2A). First, the accuracy of the mass spectrometer was normal after measurement (Figure 2B) because most of the spectra have a first-order mass error within 10 ppm. This is also consistent with the law of tryptic hydrolysis because they are mostly present in a certain number of amino acids (about 7–20). The length of the peptide meets the requirements of mass spectrometry identification (Figure 2C). We calculated and counted the number of phosphorylation sites on proteins one by one. The protein with only one phosphorylation site accounted for 41.8% of all phosphorylated proteins. The SRRM2 protein had 222 phosphorylation sites and was the protein with the most phosphorylation sites (Figure 2D). The PCA (principal component analysis) diagrams of quantitative proteomics and quantitative phospho-proteomics show that each treatment group (MDCK cells treated with ETX for 0, 10, and 30 min) was clustered within the group and significantly separated between the groups. It indicated that the biological repeatability within each treatment group was good; meanwhile, there were obvious differences between groups, and it was valuable to mine differential proteins and phosphorylated proteins (Figure 2E,F).

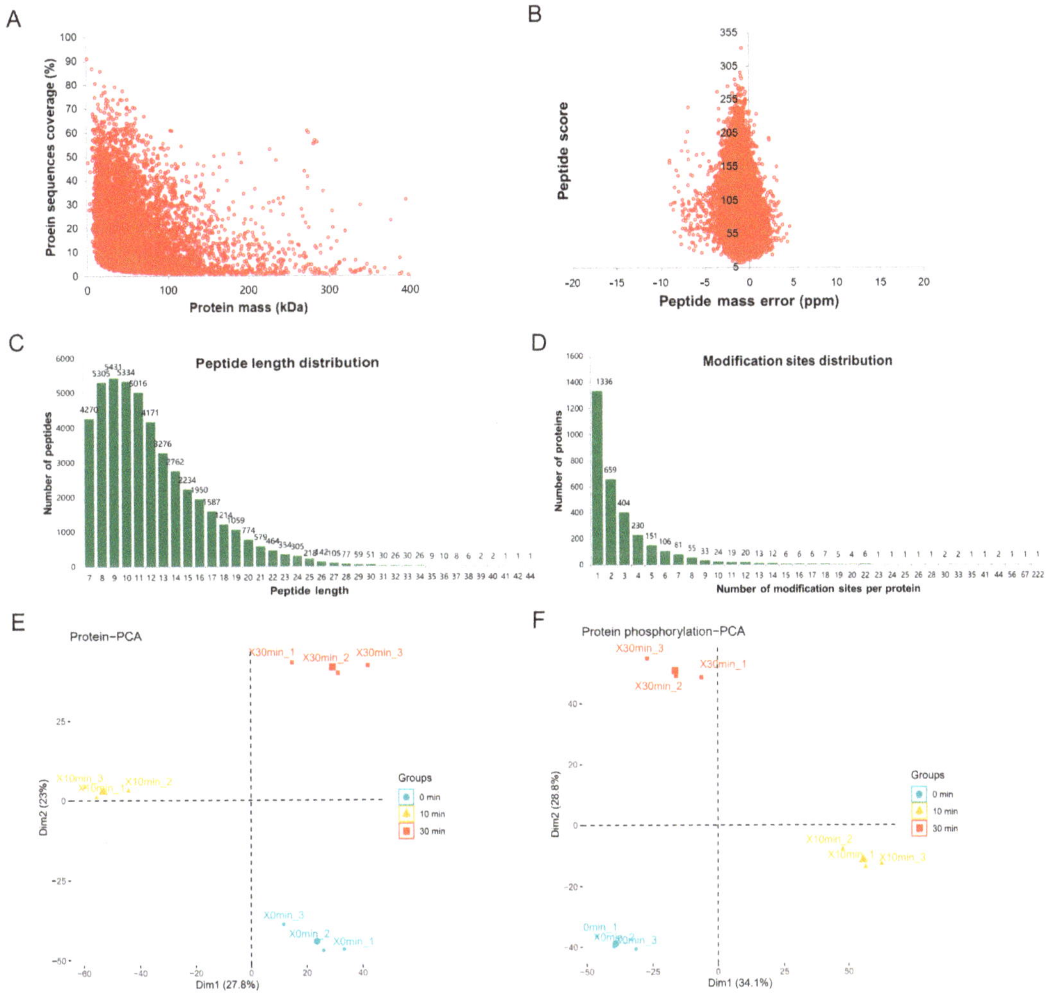


**Figure 2.** Quality control of identified proteins and phosphorylated proteins. (**A**) Relationship between protein coverage and molecular weight. (**B**) Mass accuracy distribution of mass spectrometer. (**C**) Length distribution of all identified peptides. (**D**) Distribution of phosphorylated proteins based on the number of phosphorylation sites. PCA diagram of quantitative proteomic data (**E**) and phosphorylated proteomic data (**F**).

Compared with the untreated group, a total of 44 DEPs were screened out in the 10 min ETX treatment group, including 9 up-regulated proteins and 35 down-regulated proteins (Figure 3A). A total of 588 DPPs were screened out, including 398 proteins that were up-regulated and 190 proteins that were down-regulated (Figure 3B). Compared with the untreated group, a total of 73 DEPs were screened in the 30 min ETX treatment group, including 36 up-regulated proteins and 37 down-regulated proteins (Figure 3A), and a total of 489 DPPs were screened, including 313 up-regulated proteins and 176 down-regulated proteins (Figure 3B). All up- and down-regulated DPPs and DEPs are listed (Table S3). There were 17 DEPs (Table 1) and 357 differential phosphorylation sites in the intersection of the 10 min group and the 30 min group (the phosphorylation sites up- and down-regulated in the top five are shown in Table 2). From the quantitative heat map of each treatment time, it can be seen that different clustered proteins had different protein expression and modification patterns. For example, some proteins and phosphorylated proteins were continuously up-regulated with the increase of treatment time, while others were continuously down-regulated (Figure 3C,D).

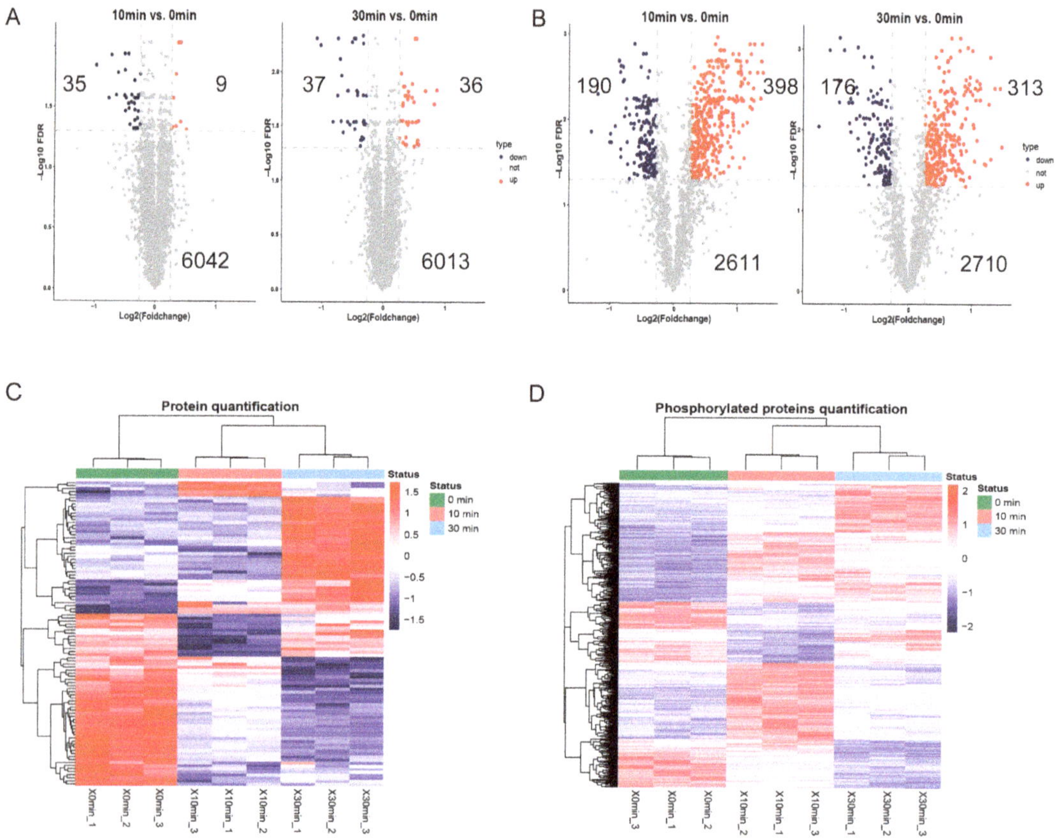


**Figure 3.** Identification of differentially expressed proteins and differentially phosphorylated proteins. (**A**) Volcano map of quantitative proteome. (**B**) Volcano map of quantitative phosphoproteome. (**C**) Heat map of three quantitative protein groups. (**D**) Heat map of three quantitative phosphorylated protein groups.

**Table 1.** Common differentially expressed proteins in 10 min group and 30 min group.

| Protein Accession | Protein Description | Gene me | FC10 | FC30 | FDR10 | FDR30 | Type |
|---|---|---|---|---|---|---|---|
| E2RED7 | Bromodomain adjacent to zinc finger domain 1B | BAZ1B | 1.328 | 1.33 | 0.009233 | 0.01666 | up |
| J9PAY8 | DNA topoisomerase 2 | TOP2A | 1.295 | 1.257 | 0.009233 | 0.013718 | up |
| E2R5T9 | Actinin alpha 4 | ACTN4 | 0.727 | 0.665 | 0.011489 | 0.015039 | down |
| F1PLS4 | Vimentin | VIM | 0.697 | 0.595 | 0.011489 | 0.016655 | down |
| E2RNB0 | Histone H2A | H2AX | 0.505 | 0.471 | 0.014266 | 0.004933 | down |
| L7N0L3 | Histone H4 | H4C7 | 0.699 | 0.598 | 0.015655 | 0.004933 | down |

**Table 1.** *Cont.*

| Protein Accession | Protein Description | Gene me | FC10 | FC30 | FDR10 | FDR30 | Type |
|---|---|---|---|---|---|---|---|
| E2RR73 | Histone cluster 1 H1 family member a | HIST1H1A | 0.647 | 0.492 | 0.016482 | 0.005695 | down |
| E2QY07 | Actinin alpha 1 | ACTN1 | 0.831 | 0.727 | 0.016911 | 0.004933 | down |
| F1Q0N9 | Keratin 19 | KRT19 | 0.753 | 0.699 | 0.025294 | 0.005341 | down |
| F1Q272 | WAS/WASL interacting protein family member 2 | WIPF2 | 0.633 | 0.665 | 0.025294 | 0.028942 | down |
| F1PB61 | TATA-box binding protein associated factor 15 | TAF15 | 0.812 | 0.713 | 0.025522 | 0.029035 | down |
| F1PRB0 | IF rod domain-containing protein | KRT86 | 0.696 | 0.622 | 0.025522 | 0.010853 | down |
| F1PW86 | Inositol 1,4,5-trisphosphate receptor type 1 | ITPR1 | 0.778 | 0.721 | 0.026721 | 0.033783 | down |
| E2REU6 | F rod domain-containing protein | KRT18 | 0.751 | 0.669 | 0.028972 | 0.015039 | down |
| E2RJ54 | Heterogeneous nuclear ribonucleoprotein K | HNRNPK | 0.727 | 0.792 | 0.029708 | 0.029035 | down |
| L7N071 | Actinin alpha 4 | ACTN4 | 0.777 | 0.732 | 0.029708 | 0.028942 | down |
| E2R6 | PEST proteolytic sigl containing nuclear protein | PCNP | 0.703 | 0.612 | 0.034266 | 0.007628 | down |

Protein accession: protein accession number of database used for search, Canis lupus familiaris (Dog) database (ID: 9615). Protein description: protein functional description. Gene me: indicates the name of the gene that code for the protein sequence. FC10: fold change of 10 min. FC30: fold change of 30 min. FDR10: false discovery rate of 10 min. FDR30: false discovery rate of 30 min.

**Table 2.** Common Top5 up-regulated and Top5 down-regulated protein phosphorylation site in 10 min group and 30 min group.

| Protein Accession | Protein Description | Gene me | Phosphosites | FC10 | FC30 | FDR10 | FDR30 | Type |
|---|---|---|---|---|---|---|---|---|
| F1PLN6 | Tripartite motif containing 33 | TRIM33 | S987 | 0.181 | 0.104 | 0.006239 | 0.03292 | down |
| F1PU14 | PDGFA associated protein 1 | PDAP1 | Y71 | 0.202 | 0.252 | 0.00203 | 0.002436 | down |
| F6XZH6 | Ladinin 1 | LAD1 | S462 | 0.503 | 0.271 | 0.014744 | 0.001249 | down |
| F1Q2Z4 | Serine/arginine repetitive matrix 2 | SRRM2 | S2394 | 0.728 | 0.327 | 0.017377 | 0.003296 | down |
| F1PKT2 | Lipolysis stimulated lipoprotein receptor | LSR | S592 | 0.437 | 0.333 | 0.005074 | 0.001034 | down |
| J9P969 | AHK nucleoprotein | AHK | S4691 | 3.607 | 3.563 | 0.003419 | 0.007272 | up |
| E2RR55 | Nuclear factor of activated T cells 2 interacting protein | NFATC2IP | S90 | 2.512 | 3.230 | 0.03395 | 0.000727 | up |
| J9NRJ1 | 40S ribosomal protein S6 | --- | S235 | 2.604 | 3.169 | 0.001298 | 0.001109 | up |
| E2RJ15 | Myotubularin related protein 10 | MTMR10 | S600 | 4.211 | 3.095 | 0.001298 | 0.003654 | up |
| J9NRJ1 | 40S ribosomal protein S6 | --- | S236 | 2.278 | 3.077 | 0.00203 | 0.000994 | up |

Protein accession: protein accession number of database used for search, Canis lupus familiaris (Dog) database (ID: 9615). Protein.description: protein functional description. Gene me: indicates the name of the gene that code for the protein sequence. FC10: fold change of 10 min. FC30: fold change of 30 min. FDR10: false discovery rate of 10 min. FDR30: false discovery rate of 30 min. ---: the gene name is not currently available in the database.

## 2.3. Subcellular Localization and KEGG Enrichment Analysis

We counted the percentage of subcellular localization of DEPs and DPPs. Eighty percent of the DEPs in the 10 min and 30 min groups were localized in the nucleus, mitochondria, and cytoplasm (Figure S1A). Eighty percent of the DPPs in the two groups were located in the nucleus and cytoplasm, and the proportion of mitochondria decreased (Figure S1B).

In order to explore the pathway involved in the toxicity of ETX, gene set enrichment analysis (GSEA) of KEGG was performed on proteomic and phosphorylated proteomic data. The five enriched pathways with the most significant up-regulated trends in the 10 min group of proteomic data were "thermogenesis", "GABAergic synapse", "amyotrophic lateral sclerosis", "oxidative phosphorylation", and "mineral absorption." The two pathways with the most significant down-regulated trends were "coronavirus disease" and "ribosome biogenesis in eukaryotes." The five enriched pathways with the most significant up-regulated trends in the 30 min group of proteomic data were "carbon metabolism", "Glycolysis/Gluconeogenesis", "endocrine and other factor-regulated calcium reabsorption", "fatty acid degradation", and "endocytosis". The three pathways with the most significant down-regulated trends were "ribosome biogenesis in eukaryotes", "ribosome", and "RNA transport" (Figure 4A).

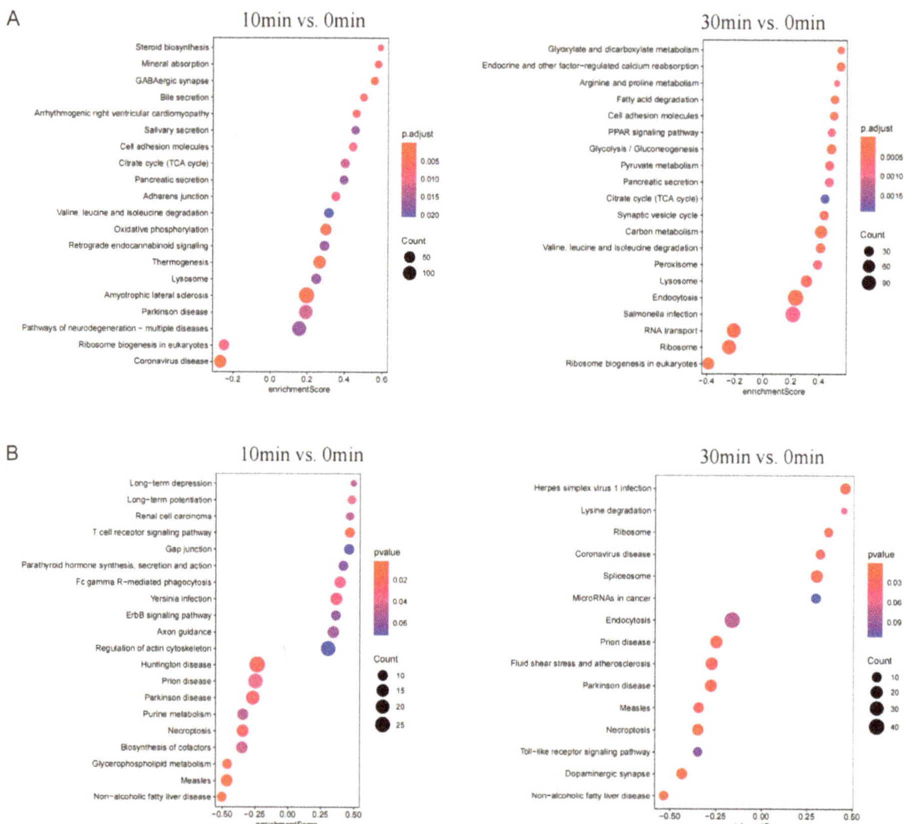


**Figure 4.** GSEA KEGG enrichment analysis. (**A**) Bubble chart of KEGG results of proteomics. (**B**) Bubble chart of KEGG results of phospho-proteomics. If more than 20 pathways are significantly enriched, the first 20 are displayed. Instead, all enrichment pathways are shown.

The five enriched pathways with the most significant up-regulated trends in the 10 min group of phosphorylated proteomic data were "T cell receptor signaling pathway", "Yersinia infection", "Fc gamma R-mediated phagocytosis", "long-term potentiation", and "renal cell carcinoma." The five pathways with the most significant down-regulated trends were "measles", "non-alcoholic fatty liver disease", "glycerophospholipid metabolism", "Huntington's disease", and "necroptosis." The five enriched pathways with the most significant up-regulated trends in the 30 min group of phosphorylated proteomic data were "Herpes simplex virus 1 infection", "spliceosome", "ribosome", "coronavirus disease", and "lysine degradation." The five pathways with the most significant down-regulated trends were "non-alcoholic fatty liver disease", "dopaminergic synapse", "necroptosis", "Parkinson disease", and "prion disease" (Figure 4B). The specific data are shown in Table S4.

### 2.4. Motif Analysis of Phosphorylation Sites

We used Motif-X to obtain the sequence characteristics around ETX-related phosphorylation sites and calculated the frequency of amino acid residues on both sides of typical phosphorylation sites (Ser/Thr sites). Taking threonine as the phosphorylation site, the frequency of Asp, Glu, Lys, Pro, and Arg in the surrounding amino acid residues was high, and the motif of Pro and Asp was highly enriched at the position of + 1 (Figure 5A, Table S5). With serine as the phosphorylation site, the frequency of Lys, Pro, and Arg in the surrounding amino acid residues was high. At the same time, the motif of Pro was highly enriched at the position of + 1 (Figure 4B, Table S5).

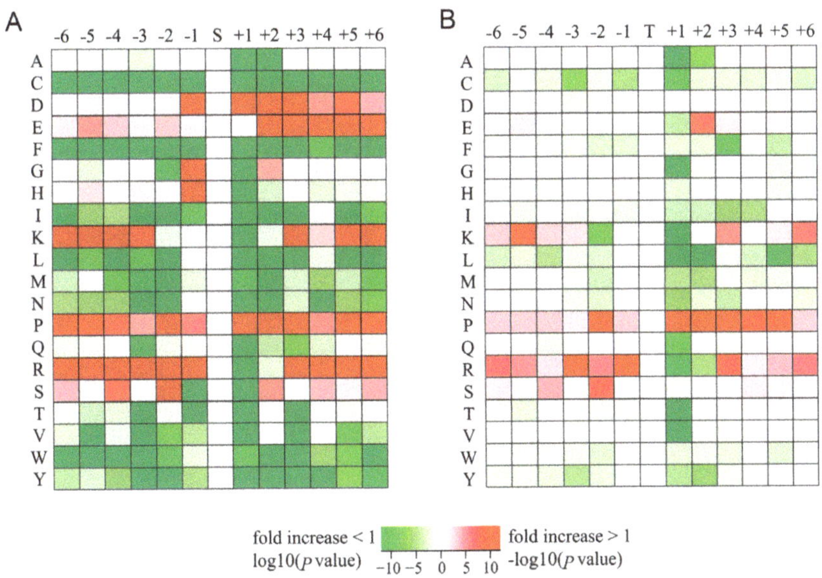


**Figure 5.** The distribution map of amino acid residues flanking the phosphorylation sites. (**A**) Distribution of amino acid residues with serine as phosphorylation site. The number of phosphorylated proteins considered was 6028. (**B**) Distribution of amino acid residues with threonine as phosphorylation site. The number of phosphorylated proteins considered was 619.

### 2.5. Integration Analysis of Differentially Expressed Proteins and Differentially Phosphorylated Proteins

In order to more comprehensively identify the toxic mechanism of ETX, we integrated the DEPs and DPPs of the 10 min group and the 30 min group into GO, KEGG enrichment analysis, and PPI network construction.

There were 565 DEPs and DPPs in the 10 min group and 491 DEPs and DPPs in the 30 min group (Figure 6A). The two groups of DEPs and DPPs were enriched by GO and KEGG, respectively, and the co-enriched GO pathways were visualized. The most important biological process (BP) in the two groups of cells is "RNA processing", the most important cellular component (CC) is "ribonucleoprotein complex", and the most important molecular function (MF) is "mRNA binding" (Figure 6B). The top three co-significant pathways were "spliceosome", "focal adhesion", and "tight junction" (Figure 6C).

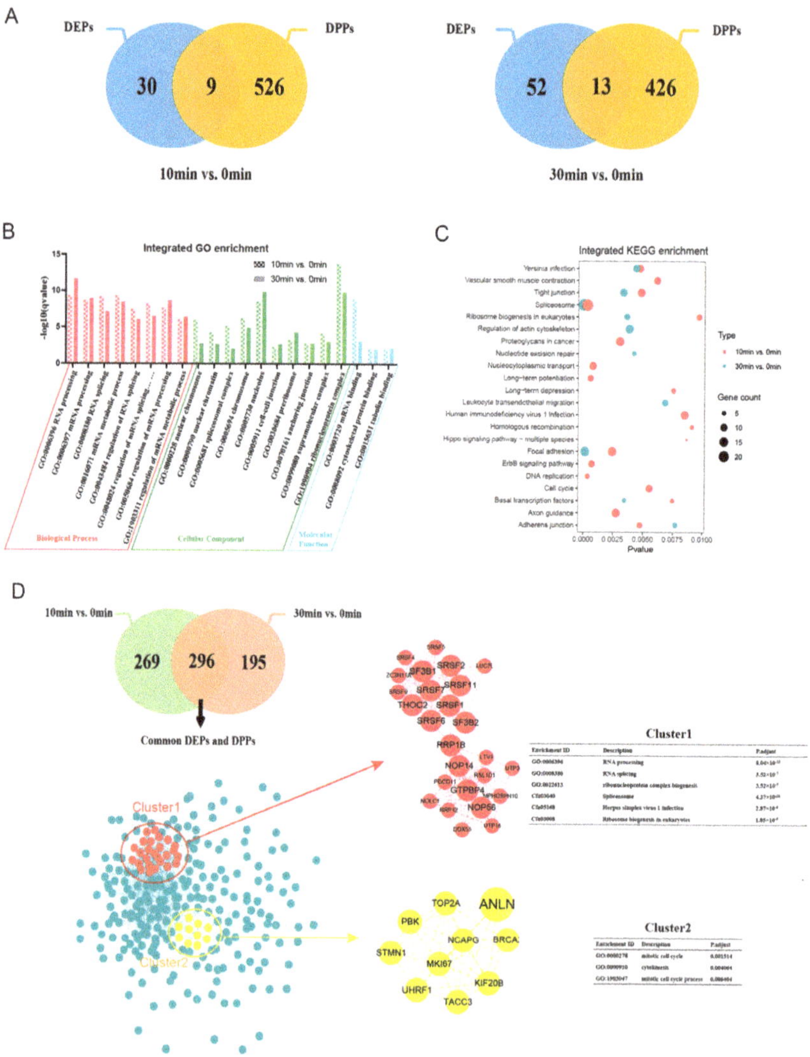


**Figure 6.** Integration analyses of differentially expressed proteins (DEPs) and differentially phosphorylated proteins (DPPs). (**A**) The union of DEPs and DPPs in the two groups. (**B**) Go terms enriched by the two groups. (**C**) KEGG pathways enriched by the two groups. (**D**) PPI network of DEPS/DPPS between the two groups. The big circle in the sub-network represents the top 20 degrees proteins in the total network. DEPs: differentially expressed proteins; DPPs: differentially phosphorylated proteins.

There were 296 DEPs and DPPs in the intersection of the 10 min group and the 30 min group. The PPI (protein–protein interaction) network of intersection proteins was constructed, and the extent of each protein in the network was calculated using Cytoscape (version 3.8.2) software's Cytotumor plug-in to screen the top 20 hub proteins (Table S6). Using the MCODE plug-in, two sub-network clusters were selected from the complex network. Cluster1 had 26 points and 140 edges and contained the 12 proteins of the top 20 hub proteins ranked by degree in the total networks; its main biological processes are RNA processing and splicesome. Cluster2 had 10 points and 36 edges, including 1 protein (ANLN) of the top 20 proteins ranked by degree in the total network. The biological process it mainly participates in is related to the cell cycle (Figure 6D).

### 2.6. Inhibitors Reduced Toxicity of ETX on MDCK Cells

After analyzing the 13 homologous key proteins of the KEGG pathway one by one in the cluster, 9 proteins could be found corresponding to the pathway (Table 3). To demonstrate whether these proteins are key to determining that ETX induced the death of MDCK cells, we conducted experiments to verify it.

**Table 3.** Analysis and arrangement of KEGG pathways of 13 homologous key proteins in the cluster in 10 min group and 30 min group.

| Gene me | Protein Accession | Type10 | Type30 | Subcellular Localization | KEGG Pathway |
|---|---|---|---|---|---|
| SRSF1 | E2RJL3 | down | down | nucleus | cfa04657; cfa05168; cfa03040 |
| SRSF2 | E2RDB2 | up | up | nucleus | cfa05168; cfa03040 |
| SRSF6 | F1PTE0 | up | up | nucleus | cfa05168; cfa03040 |
| SRSF7 | J9P2I8 | down | up | nucleus | cfa05168; cfa03040 |
| SF3B1 | F1PKF6 | up | up | plasma membrane | cfa03040 |
| SF3B2 | E2RL65 | up | up | nucleus | cfa03040 |
| GTPBP4 | F1PJY9 | up | up | nucleus | cfa03008 |
| THOC2 | J9PB31 | down | down | cytoplasm | cfa03013; cfa03040 |
| NOP56 | E2QU53 | up | up | nucleus | cfa03008 |
| NOP14 | F1PI93 | up | up | plasma membrane | 0 |
| ANLN | F1PZ90 | up | up | nucleus | 0 |
| SRSF11 | F1P6U8 | down | down | nucleus | 0 |
| RRP1B | F1PS87 | down | down | cytoplasm, nucleus | 0 |

Protein accession: protein accession number of database used for search. Gene me: indicates the name of the gene that code for the protein sequence. KEGG: pathways of proteins in KEGG database.

We found that these nine proteins were involved in three pathways, most of which were concentrated in the SR protein complex of the spliceosome pathway. Therefore, we selected SRPK1, a common upstream protein of this pathway, for verification. We used an inhibitor, SPHINX31, that has been shown to be effective in inhibiting the phosphorylation of serine/arginine-rich splicing factor 1 (SRSF1) and is also a selective inhibitor of serine/arginine-rich protein kinase 1 (SRPK1) [22,23]. MDCK cells were treated with different concentrations of SPHINX31 for 12, 24, 36, and 48 h before being exposed to ETX (8.93 nM), and it was found that the survival rate of MDCK cells was enhanced in a concentration-dependent manner (Figure 7A–D). In addition, SRPIN340 is also an ATP-competing SRPK inhibitor and has been shown to effectively inhibit SRPK1 expression. Similarly, different concentrations of SRPIN340 were applied to MDCK cells, and it can be seen that SRPIN340 could effectively inhibit the death of MDCK cells at low concentrations but had certain cytotoxicity when the concentration was greater than 80 μM (Figure 7E–H). The results suggest that SRSF1 and SRPK1 are involved in ETX-induced toxicity on MDCK cells.

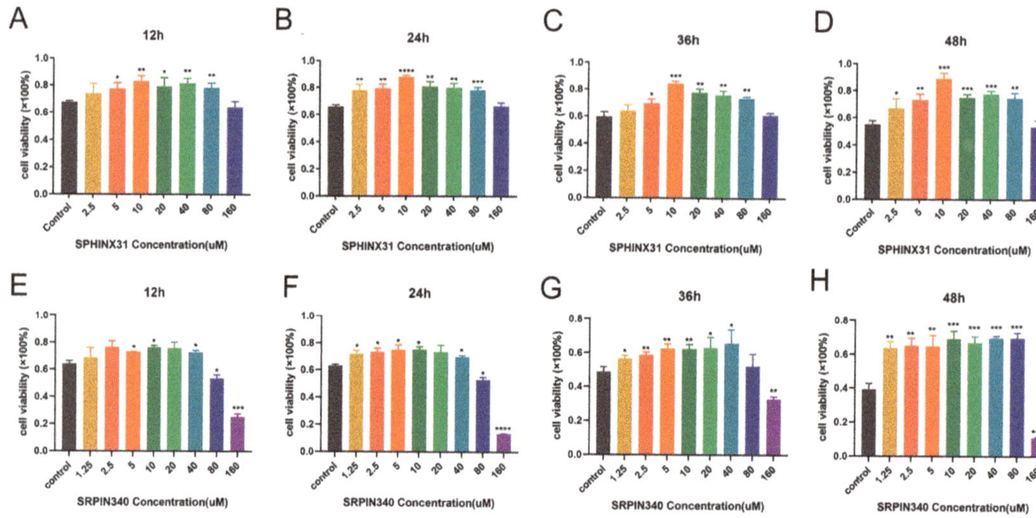


**Figure 7.** Inhibiting the expression of SRSF1 and SRPK1 can effectively inhibit the lethality of ETX on MDCK cells. (**A–D**) MDCK cells were treated with different concentrations of SPHINX31 for 12, 24, 36, and 48 h, and the inhibitor can improve the survival rate of ETX after attacking MDCK cells. (**E–H**) MDCK cells were treated with different concentrations of SRPIN340 for 12, 24, 36, and 48 h, which also improves survival after ETX attacks MDCK cells. Values are the mean ± SD ($n$ = 3). At least three independent biological replicates were performed for each experiment. * $p < 0.05$, ** $p < 0.01$, *** $p < 0.001$, and **** $p < 0.0001$; this was compared to controls in each group that did not use lentivirus.

### 2.7. SRSF1 and SRPK1 Inhibit the Toxicity of ETX to MDCK Cells

In addition, to further identify the central proteins of MDCK that caused death by ETX, we selected six key proteins on the three key pathways obtained, NOP56, SF3B1, SF3B2, THOC2, SRSF1 and SRPK1, and designed lentivirus to knock down the expression levels of these proteins in MDCK cells, respectively (information on customized lentiviruses is provided in Table S7). The results expressed that ETX could lead to the increase of SRPK1 in MDCK cells, and the use of the four lentiviruses could reduce the expression of SRPK1 in MDCK cells (Figure 8A). In addition, the levels of these six key proteins in MDCK cells were also knocked down by corresponding lentiviruses (Figure 8B). The results of the cell cytotoxicity assay showed that the specific knockdown of SRSF1 and SRPK1 by lentivirus could effectively inhibit the cell death of MDCK cells (Figure 8C), which was consistent with the results of previous inhibitors. However, NOP56 does not play a role in cell death, and decreased SF3B1 expression can promote the death of MDCK cells (Figure 8D,E).

Additionally, to further confirm our conclusion, we selected two key proteins in the spliceosome pathway, SF3B2 and THOC2, and constructed lentiviruses to knock down their expression levels in MDCK cells. The results showed that SF3B2 could inhibit the toxicity of ETX to MDCK cells, while THOC2 could inhibit cell death at ETX = 1.79 nM, but the inhibitory effect was not significant at high concentrations of ETX (Figure 8F). The negative control of the lentivirus vector used had no effect on cell viability (Figure 8G). These results support our findings in the inhibitor experiments above that SRSF1, SRPK1, and their spliceosome pathways play critical roles in ETX-induced cytotoxicity in MDCK cells (Figure 7).

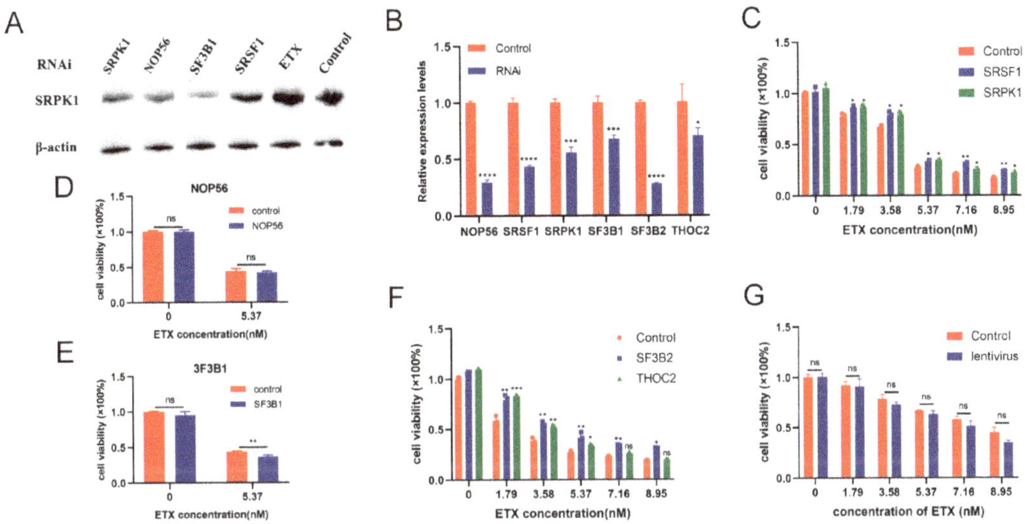


**Figure 8.** Lentiviruses using SRSF1 and SRPK1 can effectively inhibit the toxicity of ETX to MDCK cells. (**A**) ETX can increase the expression of SRPK1 protein in MDCK cells, and the four lentiviruses we synthesized can reduce the expression of SRPK1. (**B**) The relative quantitative results of RT-qPCR showed that the six lentiviruses could significantly reduce the expression levels of corresponding proteins in MDCK cells. (**C**) Lentivirus-decreased expressions of SRSF1 and SRPK1 could effectively inhibit the death of MDCK cells. (**D**,**E**) NOP56 does not play a role in cell death, and decreased SF3B1 expression can promote the death of MDCK cells. (**F**) Knockdown of SF3B2 and THOC2 could reduce the death rate of MDCK cells caused by ETX to a certain extent. (**G**) The lentiviral vector did not affect the survival of the cells. The above experimental cells were all exposed to ETX for 1 h. Values are the mean ± SD ($n = 3$). At least three independent biological replicates were performed for each experiment. * $p < 0.05$, ** $p < 0.01$, *** $p < 0.001$, and **** $p < 0.0001$; this was compared to controls in each group that did not use lentivirus.

## 3. Discussion

ETX is highly toxic and can cause disease and even death in a variety of animals [24]. Because of this, it is considered a potential agent of biological and toxin warfare. After ETX poisoning, the course of the disease develops rapidly, and the mortality rate is high [25]. Therefore, it is necessary to reveal the pathogenesis of ETX, search for key signaling pathways and molecules, and strive to provide help and therapeutic targets for subsequent therapeutic research.

In the enriched ETX-related pathways, there were many transcription- and translation-related words, such as “RNA transport”, “ribosome”, “splicesome”, “RNA processing”, “mRNA metabolic process”, etc., and most of the DEPs and DPPs were located in the nucleus, cytoplasm, and mitochondria. At the same time, the core sub-network in the PPI network was also enriched in “RNA processing”, “RNA spreading”, and other processes, indicating that the toxicity of ETX may be closely related to interfering with the transcription and translation of host cells. In order to verify the role of these hub proteins in ETX poisoning, we systematically summarized the pathways where these proteins were located, and they were found to be almost concentrated in three pathways: Herpes simplex virus 1 infection and spliceosome and ribosome biogenesis in eukaryotes. What is more interesting is that six of these proteins (SF3B1/2, SRSF1/2/6, and THOC2) exist in the “spliceosome” pathway. The spliceosome, which removes introns from genes and connects exons together to synthesize pre-mRNA, plays a crucial role in gene expression [26,27].

Abnormal RNA splicing causes a variety of diseases [28]. Interestingly, SF3B1/2 and SRSF1/2/6 are both present in the SR protein complex of this pathway; SR proteins promote the recruitment of spliceosome components through protein–protein interactions to promote early spliceosome assembly, and these functions are regulated by reversible phosphorylation [29]. Therefore, we suspect that these proteins may be the key proteins that determine ETX poisoning. Studies have proved that high expression of splicing factor SRSF1 can promote the occurrence and development of lung cancer [30], breast cancer [31], pancreatitis [32], and other diseases and cancers, indicating that SRSF1 is a key protein in the regulation of cancer cell expression. Studies have shown that the abnormal expression of the SRSF2 gene is strongly correlated with acute myeloid leukemia (AML) [33], liver cancer [22,34], lung cancer [35], and other diseases. Recent studies have found that HSV-1 (herpes simplex virus-1) infection can up-regulate SRSF2 levels, and SRSF2 can negatively regulate HSV-1 transcription [36]. Therefore, SRSF2 is expected to be a therapeutic target for a variety of diseases. Multiple alternative splicing of SRSF6 also provides therapeutic targets for diseases such as rectal cancer [37], gastric cancer [38], and alcoholic liver disease [39]. SF3B1 is an important component of the spliceosome. Reduced expression of this gene can reduce the incidence of many human cancers and can also slow the progression of cancer [40–42]. But we have only found a finished inhibitor of the SRSF1 protein, SPHINX31, which can inhibit the phosphorylation of SRSF1, and no specific antibodies against them have been found.

We then went further up the pathway and found that the common upstream protein of these proteins was SRPK1. SRPK1 is a protein that can be involved in mRNA processing, including, of course, the alternative splicing mentioned earlier [43]. One of the unique features of SRPK1 is that it phosphorylates proteins containing the SR domain. Studies have shown that the high expression of SRPK1/2 is the cause of enhanced phosphorylation and nuclear translocation of SRSF1 [44]. Therefore, we have reason to believe that SRPK1 plays a key role in mediating ETX-induced MDCK cell poisoning. Interestingly, we found that SRPIN340 is an ATP-competitive SRPK inhibitor that inhibits SRPK1 activity. Therefore, in our results, we found that the reduction of SRSF1 and SRPK1 activity can effectively inhibit the death of MDCK cells induced by ETX. Lentivirus knockdowns of the two proteins produced the same results, and the increase in cell survival was more pronounced in the high-concentration ETX group.

Based on the above results, we suspect that when ETX acts on MDCK cells, SRPK1-specific kinase phosphorylates multiple proteins in the SR protein complex, thereby inhibiting the normal splicing of mRNA in the cells and preventing the synthesis of correct proteins, resulting in cell death. As has been reported in many studies, our study also demonstrates that SRPK1 does affect the normal life cycle of cells. However, it plays an important part in cancer treatment and diagnosis [45], whether it helps in the treatment of diseases caused by ETX is unknown, and this is a question that we must continue to address. This also suggests that ETX-induced disease and injury are not as simple as the formation of pores in the membrane but have more complex damage mechanisms, and our findings are an important starting point.

Unfortunately, of the 13 proteins we screened, most currently have no finished inhibitors and antibodies, and several proteins are located in unknown pathways, making our research and validation extremely difficult. Further exploration of these unknown proteins is essential. The research on the pathogenesis of ETX needs to be ongoing. The above differential proteins can be used as potential therapeutic sites for ETX intoxication.

## 4. Conclusions

Taken together, we used quantitative proteomics and quantitative phospho-proteomics techniques to screen ETX pathogenic-related proteins, phosphorylated proteins, and some important pathways in the MDCK cell line. We identified the key proteins SRPK1, SF3B2,

THOC2, and SRSF1 in ETX-induced cell death and verified that their high expression can lead to abnormal splicing of MDCK cells, leading to cell death. To our knowledge, this is the first joint proteomic and phosphorylation proteomic study of ETX on MDCK cells, which provides data support for clarifying the pathogenic mechanism, screening toxic biomarkers, and looking for therapeutic targets of ETX intoxication.

## 5. Materials and Methods

### *5.1. Recombinant Toxin and Reagents*

SRPK1 Polyclonal Antibody was purchased from Immunoway (Plano, TX, USA). Horseradish peroxidase (HRP)-conjugated goat anti-rabbit IgG (H + L) antibody was purchased from Solarbio (Beijing, China). SPHINX31 (HY117661) and SRPIN340 (HY13949) were purchased from Med Chem Express Corporation (MCE, Romulus, MI, USA). SRSF1, SRPK1, SF3B1, and NOP56 four RNAi-custom lentiviruses were built by GENE Corporation (Shanghai, China). RIPA Lysis Buffer abs929 (strong) was purchased from Absin (Shanghai, China).

Recombinant toxins, GST-labeled ETX, were expressed and purified according to the basic procedures previously performed in the laboratory [14,46].

### *5.2. Cell Culture*

MDCK cells were cultured in DMEM complete medium (Dulbecco's Modified Eagle Medium, Gibco, Carlsbad, CA, USA), and the conditions of the incubator were set at 37 °C with 5% $CO_2$.

### *5.3. Western Blotting*

To assess the effects of ETX (0.9 nM) on MDCK cells, $1 \times 10^5$/mL cells were plated in 150 mm plates for 24 h. Subsequently, the cells were incubated with ETX (0.9 nM) in DMEM for different durations (0, 0.5, 5, 10, 15, 20, 25, and 30 min) at 37 °C. After incubation, cells were washed three times with PBS. According to the manufacturer's instructions (absin, Shanghai, China, abs9229), 25 μL of PMSF at a concentration of 100mM was added to 225 μL of RIPA lysate to make the final concentration of PMSF 1 mM. After mixing evenly, 250 μL of the mixture was added to each well, and the cells were lysed for 1 min. After which the cells attached to the bottom of the plate were scraped off with a scraper and collected into a centrifuge tube; this procedure was performed on ice. The lysates were then centrifuged at 14,000× *g* for 5 min at 4 °C. The supernatants obtained by centrifugation were collected for subsequent experiments. Samples were mixed with loading buffer and added to a 4–20% polyacrylamide SDS-PAGE gel for electrophoresis, after which they were transferred to PVDF membranes. The membranes were blocked with 5% skim milk powder for approximately 3 h at room temperature by gentle shaking and washed three times using PBST. Then, the membranes were incubated with anti-SRPK1 polyclonal overnight at 4 °C and washed three times using PBST. The membranes were incubated with goat anti-rabbit IgG secondary antibody for 1 h. Finally, the color development solution was added, and the results were analyzed and imaged using an AE-1000 cool CCD image analyzer.

The Western blotting results were quantified using image J software (version number: Java 1.8.0-112).

### *5.4. Protein Extraction*

MDCK cells treated with ETX for 10 min and 30 min were collected, then lysate (8 M urea, 1% protease inhibitor) was added and crushed using ultrasonic treatment on ice for 10 s. The crushed liquid was centrifuged at 4 °C, 14,000× *g*, for 10 min. The supernatant was collected, and the protein concentration was determined.

### *5.5. Trypsin Digestion*

The sample pretreatment before digestion was as follows [47]: The protein was reduced with 5 mM dithiothreitol for 30 min and then alkylated with 11 mM iodoacetamide under dark conditions for 15 min. Finally, the urea concentration of the sample was diluted to less than 2 M by adding 100 mM $NH_4HCO_3$. Then, the protein solution was digested twice, first at a 1:50 ratio of trypsin to protein overnight, followed by a 1:100 ratio of trypsin to protein for 4 h. Finally, the peptides were desalted by Strata X SPE column.

### 5.6. TMT Labeling

Tryptic peptides were first dissolved in 0.5 M TEAB. Each channel of peptide was labeled with their respective TMT reagent (based on manufacturer's protocol, Thermo Scientific, Rockford, IL, USA, cat.90406) and incubated for 2 h at room temperature. Five microliters of each sample were pooled, desalted, and analyzed by MS to check labeling efficiency. After labeling efficiency check, samples were quenched by adding 5% hydroxylamine. The pooled samples were then desalted with Strata X SPE column (Phenomenex) and dried by vacuum centrifugation.

### 5.7. HPLC Fractionation

Separation was performed using a high pH reversed-phase HPLC on Agilent 300Extend C18 column (5 μm particles, 4.6 mm ID, 250 mm long). The 8–32% acetonitrile gradient of PH9.0 separated 60 fractions. The resulting fractions were recombined and dried, resulting in 18.

### 5.8. LC-MS/MS Analysis

The tryptic peptides were dissolved in solvent A (0.1% formic acid, 2% acetonitrile/in water), directly loaded onto a homemade reversed-phase analytical column (25 cm length, 75/100 μm i.d.). Peptides were separated with a gradient from 8% to 25% solvent B (0.1% formic acid in 90% acetonitrile) over 40 min, 25% to 35% in 14 min, and climbing to 80% in 3 min, then holding at 80% for the last 3 min, all at a constant flowrate of 700 nL/min on an EASY-nLC 1000 UPLC system (Thermo Fisher Scientific, Waltham, MA, USA).

The separated peptides were analyzed in Q ExactiveTM (Thermo Fisher Scientific, Waltham, MA, USA) with a nano-electrospray ion source. The electrospray voltage applied was 2.1 kV. The full MS scan resolution was set to 70,000 for a scan range of 350–1800 $m/z$. Up to 10 most abundant precursors were then selected for further MS/MS analyses with 15 s dynamic exclusion. The HCD fragmentation was performed at a normalized collision energy (NCE) of 28%. The fragments were detected in the Orbitrap at a resolution of 35,000. Fixed first mass was set as 100 $m/z$. Automatic gain control (AGC) target was set at 5E4, with an intensity threshold of 5E3 and a maximum injection time of 200 ms.

### 5.9. Database Search

The resulting MS/MS data were processed using MaxQuant search engine (v.1.5.2.8). Tandem mass spectra were searched against the canis SwissProt database (25,492 entries) concatenated with reverse decoy database. Trypsin/P was specified as cleavage enzyme allowing up to 2 missing cleavages. The mass tolerance for precursor ions was set as 20 ppm in First search and 5 ppm in Main search, and the mass tolerance for fragment ions was set as 0.02 Da. Carbamidomethyl on Cys was specified as fixed modification. Acetylation on protein N-terminal, oxidation on Met, and deamidation (NQ) were specified as variable modifications. TMT-10plex quantification was performed. FDR was adjusted to <1% and minimum score for peptides was set to >40.

### 5.10. Bio-Material-Based PTM Enrichment (For Phosphorylation)

Enrichment was performed using IMAC kit (ThermoFisher Scientific, Rockford, IL, USA, Cat. A32992) as follows. Peptide mixtures were first incubated with IMAC microspheres suspension with vibration in loading buffer (50% acetonitrile/0.5% acetic acid). To remove the non-specifically adsorbed peptides, the IMAC microspheres were washed with 50% acetonitrile/0.5% acetic acid and 30% acetonitrile/0.1% trifluoroacetic acid, sequentially. To elute the enriched phosphopeptides, the elution buffer containing 10% $NH_4OH$

was added, and the enriched phosphopeptides were eluted with vibration. The supernatant containing phosphopeptides was collected and lyophilized for LC-MS/MS analysis.

### 5.11. Standardization of Data

First, the anti-library and the contaminated library were removed. Then, the data were first laterally homogenized (replacing 0 values with NaN and dividing the intensity values of each protein in all samples by the mean of the intensity values of all proteins in all samples) and then longitudinally normalized (dividing each column by the median of that column). Finally, unique peptides were grouped (peptide segments belonging to the same protein are grouped together), and the median of quantitative values in each group of peptides was used as the quantitative value of the protein.

### 5.12. Bioinformatics Analysis

We performed principal component analysis (PCA) and visualization using the "FactoMineR" package and "factoextra" package of R software (version 4.0.3). To screen for DEPs and DPPs, we used an independent sample *t*-test. We then applied the Benjamin and Hochberg method to correct the *p*-value for multiple testing and to obtain the false discovery rate (FDR) value. The screening criteria were a fold change greater than 1.2 or less than 0.83, and an FDR less than 0.05. We used Wolfpsort (https://wolfpsort.hgc.jp/) to predict the subcellular structural localization of these proteins. GO and KEGG enrichment analyses were performed using the "clusterProfiler" package of R software (version R-3.5.1). Motif analysis was conducted using soft motif X. We imported the protein list into the STRING (accessed on 12 July 2021) (https://www.string-db.org/) to export the relationship pairs and visualize the protein–protein interaction (PPI) network using Cytoscape software (version 3.8.2). We calculated the degree of each protein in the network using the CytoHubba. A higher degree value indicates a more central position for the protein in the network. We used the MCODE plug-in to filter the core sub-network, with a K-Core set to 2.

### 5.13. Cytotoxicity Assay

In this experiment, we used two inhibitors, SPHINX31 and SRPIN340. MDCK cells were grown in 96-well plates at a density of $5 \times 10^4$ cells/mL for 16 h. The cells were then washed two or three times with PBS, and different concentrations of SPHINX31 and SRPIN340 (ranging from 2.5 to 160 μM, diluted by double volume) were added to the wells. The control group received only medium, and three replicate wells were set for each concentration. The cells were incubated at 37 °C for 12, 24, 36, or 48 h. After the incubation period, 5.37 nM GST-ETX was added to the corresponding well and incubated in the cell incubator for 1 h. To estimate cell survival, 100 μL DMEM complete medium and 20 μL MTS (3-(4,5-dimethylthiazol-2-yl)-5(3-carboxymethoxyphenyl)-2-(4-sulfophenyl)-2H-tetrazolium inner salt) were added to the pore and absorbance at 492 nm was measured. As a positive control, cells were treated with toxin only (no inhibitor), and as a negative control, cells were treated with DMEM without the toxin and inhibitor.

### 5.14. Lentivirus-Mediated RNA Interference Assay

Six RNAi constructs were used to specifically knock down the expression levels of NOP56, SF3B1, SF3B2, THOC2, SRSF1, and SRPK1 proteins using lentivirus. MDCK cells were seeded in 96-well plates at $5 \times 10^3$ cells per well and cultured for 16 h. The cells were then gently washed two or three times with PBS. The lentivirus was diluted to a working concentration of $5 \times 10^7$ TU/mL (MOI = 100), and the lentiviral infection system (86 μL DMEM, 10 μL lentivirus, 4 μL 25 × HiTransG P) was added to the cells which were then incubated in the cell incubator for approximately 16 h. The culture was continued with DMEM culture medium and infected for about 48 h. After that, the cells were treated with different concentrations of diluted GST-ETX (0, 1.79, 3.58, 5.37, 7.16, 8.95 nM) with three replicate wells for each concentration, and incubated in the cell incubator for 1 h. Cell survival was estimated by adding 100 μL DMEM complete medium and 20 μL MTS to the wells and measuring absorbance at 492 nm. Lentivirus and toxin were not added as negative controls, while only toxin cells were added as positive control.

Western blotting was performed to determine if the lentivirus reduced protein levels. In simple terms, after infecting MDCK cells with lentivirus, total protein was extracted using RIPA lysate, separated by 4–20% SDS-PAGE at 130 V for 1 h, and transferred to a PVDF membrane. The membrane was then gently blocked with 5% skim milk at room temperature for 3 h, probed with SRPK1 rabbit antibody (1:1000, Immunoway YT4422, Plano, TX, USA), incubated overnight at 4 °C, and washed three times using PBST. The membranes were incubated with goat anti-rabbit IgG secondary antibody for 1 h. Finally, the color development solution was added, and the results were analyzed and imaged using an AE-1000 cool CCD image analyzer.

### 5.15. RT-qRCR

The knockdown efficiency of lentivirus was verified by RT-qPCR. After MDCK cells were infected with lentivirus, the cells were collected. Total RNA in cells was extracted by Trizol and then quantified using Qubit 4 Fluorometer (Invitrogen, Carlsbad, CA, USA). After treatment with *EVO M-MLV* One Step RT-qPCR Kit (SYBR) (Accurate Biotechnology, Code NO. AG11732, Hunan, China), the detection was carried out on the QuantStudio™ 7 Pro Real-Time PCR System (Thermo Fisher Scientific, Waltham, MA, USA). Relative gene expression was calculated by the $2^{-\Delta\Delta Ct}$ method [48]. Primer information is shown in Table S8.

### 5.16. Statistical Analysis

Statistical analysis was performed with GraphPadPrism9.0 software. The data were presented as the mean ± standard deviation (SD) of multiple independent experiments. Cell survival was logit transformed, and then the differences between groups were analyzed by independent sample one-way ANOVA. We prescribed that the significance level was $p < 0.05$, and the difference was considered to be statistically significant.

**Supplementary Materials:** The following supporting information can be downloaded at: https://www.mdpi.com/article/10.3390/toxins16090394/s1, Figure S1: Subcellular localization statistics; Table S1: Differential protein information; Table S2: Phosphorylation sites information; Table S3: All up- and down-regulated DPPs and DEPs' information; Table S4: All DPPs' information; Table S5: Motif-X analysis of significantly regulated phosphopeptides; Table S6: Cytohubba results Top 20 degree. Table S7: RNAi-customized-lentivirus information; Table S8: Primer information for RT-qPCR.

**Author Contributions:** Conceptualization, W.X., L.K. and N.X.; Methodology, N.Y., J.W. (Jinglin Wang) and L.K.; Software, J.H., J.W. (Jing Wang), D.L., L.D. and M.C.; Visualization, M.D.; Formal Analysis, N.Y., M.D., J.W. (Jing Wang), J.L., Y.W. and T.L.; Investigation, N.Y. and J.L.; Resources, N.X. and W.X.; Writing—original draft, N.Y., J.H., L.D. and M.C. Writing—review and editing, J.H., M.D., J.W. (Jinglin Wang), J.L., S.G., Y.W., X.L., S.H. and M.C.; Visualization, M.D. and N.Y.; Supervision, N.X. and W.X.; Project administration, N.X., L.K. and W.X.; Funding acquisition, W.X. All authors have read and agreed to the published version of the manuscript.

**Funding:** This research was funded by the National Natural Science Foundation of China (No. 31970127).

**Institutional Review Board Statement:** Not applicable.

**Informed Consent Statement:** Not applicable.

**Data Availability Statement:** The datasets generated during and/or analyzed during the current study are available from the corresponding author upon reasonable request.

**Conflicts of Interest:** The authors declare no conflicts of interest.

## References


1. Li, J.; Adams, V.; Bannam, T.L.; Miyamoto, K.; Garcia, J.P.; Uzal, F.A.; Rood, J.I.; McClane, B.A. Toxin plasmids of Clostridium perfringens. *Microbiol. Mol. Biol. Rev.* **2013**, *77*, 208–233. [CrossRef]
2. Garcia, J.P.; Giannitti, F.; Finnie, J.W.; Manavis, J.; Beingesser, J.; Adams, V.; Rood, J.I.; Uzal, F.A. Comparative neuropathology of ovine enterotoxemia produced by *Clostridium perfringens* type D wild-type strain CN1020 and its genetically modified derivatives. *Vet. Pathol.* **2015**, *52*, 465–475. [CrossRef]
3. Cases, M.; Llobet, A.; Terni, B.; de Aranda, I.G.; Blanch, M.; Doohan, B.; Revill, A.; Brown, A.M.; Blasi, J.; Solsona, C. Acute Effect of Pore-Forming *Clostridium perfringens* ε-Toxin on Compound Action Potentials of Optic Nerve of Mouse. *eNeuro* **2017**, *4*, ENEURO.0051-17.2017. [CrossRef]
4. Yao, W.; Kang, J.; Kang, L.; Gao, S.; Yang, H.; Ji, B.; Li, P.; Liu, J.; Xin, W.; Wang, J. Immunization with a novel *Clostridium perfringens* epsilon toxin mutant rETX(Y196E)-C confers strong protection in mice. *Sci. Rep.* **2016**, *6*, 24162. [CrossRef]
5. Bokori-Brown, M.; Hall, C.A.; Vance, C.; da Costa, S.P.F.; Savva, C.G.; Naylor, C.E.; Cole, A.R.; Basak, A.K.; Moss, D.S.; Titball, R.W. *Clostridium perfringens* epsilon toxin mutant Y30A-Y196A as a recombinant vaccine candidate against enterotoxemia. *Vaccine* **2014**, *32*, 2682–2687. [CrossRef]
6. Bokori-Brown, M.; Savva, C.G.; da Costa, S.P.F.; Naylor, C.E.; Basak, A.K.; Titball, R.W. Molecular basis of toxicity of *Clostridium perfringens* epsilon toxin. *FEBS J.* **2011**, *278*, 4589–4601. [CrossRef]
7. Minami, J.; Katayama, S.; Matsushita, O.; Matsushita, C.; Okabe, A. Lambda-toxin of *Clostridium perfringens* activates the precursor of epsilon-toxin by releasing its N- and C-terminal peptides. *Microbiol. Immunol.* **1997**, *41*, 527–535. [CrossRef]
8. Stiles, B.G.; Barth, G.; Barth, H.; Popoff, M.R. *Clostridium perfringens* epsilon toxin: A malevolent molecule for animals and man? *Toxins* **2013**, *5*, 2138–2160. [CrossRef]
9. Popoff, M.R. Epsilon toxin: A fascinating pore-forming toxin. *FEBS J.* **2011**, *278*, 4602–4615. [CrossRef]
10. Petit, L.; Gibert, M.; Gillet, D.; Laurent-Winter, C.; Boquet, P.; Popoff, M.R. *Clostridium perfringens* epsilon-toxin acts on MDCK cells by forming a large membrane complex. *J. Bacteriol.* **1997**, *179*, 6480–6487. [CrossRef]
11. Shortt, S.J.; Titball, R.W.; Lindsay, C.D. An assessment of the in vitro toxicology of *Clostridium perfringens* type D epsilon-toxin in human and animal cells. *Hum. Exp. Toxicol.* **2000**, *19*, 108–116. [CrossRef]
12. Ivie, S.E.; Fennessey, C.M.; Sheng, J.; Rubin, D.H.; McClain, M.S. Gene-trap mutagenesis identifies mammalian genes contributing to intoxication by *Clostridium perfringens* ε-toxin. *PLoS ONE* **2011**, *6*, e17787. [CrossRef]
13. Chassin, C.; Bens, M.; de Barry, J.; Courjaret, R.; Bossu, J.L.; Cluzeaud, F.; Mkaddem, S.B.; Gibert, M.; Poulain, B.; Popoff, M.R.; et al. Pore-forming epsilon toxin causes membrane permeabilization and rapid ATP depletion-mediated cell death in renal collecting duct cells. *Am. J. Physiol. Renal Physiol.* **2007**, *293*, F927–F937. [CrossRef]
14. Gao, J.; Xin, W.; Huang, J.; Ji, B.; Gao, S.; Chen, L.; Kang, L.; Yang, H.; Shen, X.; Zhao, B.; et al. Research articleHemolysis in human erythrocytes by *Clostridium perfringens* epsilon toxin requires activation of P2 receptors. *Virulence* **2018**, *9*, 1601–1614. [CrossRef]
15. Xin, W.; Wang, J. *Clostridium perfringens* epsilon toxin: Toxic effects and mechanisms of action. *Biosaf. Health* **2019**, *1*, 71–75. [CrossRef]
16. Savva, C.G.; Clark, A.R.; Naylor, C.E.; Popoff, M.R.; Moss, D.S.; Basak, A.K.; Titball, R.W.; Bokori-Brown, M. The pore structure of *Clostridium perfringens* epsilon toxin. *Nat. Commun.* **2019**, *10*, 2641. [CrossRef]
17. Miyata, S.; Minami, J.; Tamai, E.; Matsushita, O.; Shimamoto, S.; Okabe, A. *Clostridium perfringens* epsilon-toxin forms a heptameric pore within the detergent-insoluble microdomains of Madin-Darby canine kidney cells and rat synaptosomes. *J. Biol. Chem.* **2002**, *277*, 39463–39468. [CrossRef]
18. Petit, L.; Maier, E.; Gibert, M.; Popoff, M.R.; Benz, R. *Clostridium perfringens* epsilon toxin induces a rapid change of cell membrane permeability to ions and forms channels in artificial lipid bilayers. *J. Biol. Chem.* **2001**, *276*, 15736–15740. [CrossRef]
19. Ferrarezi, M.C.; Curci, V.C.; Cardoso, T.C. Cellular vacuolation and mitochondrial-associated factors induced by *Clostridium perfringens* epsilon toxin detected using acoustic flow cytometry. *Anaerobe* **2013**, *24*, 55–59. [CrossRef]
20. Wioland, L.; Dupont, J.L.; Doussau, F.; Gaillard, S.; Heid, F.; Isope, P.; Pauillac, S.; Popoff, M.R.; Bossu, J.L.; Poulain, B. Epsilon toxin from *Clostridium perfringens* acts on oligodendrocytes without forming pores, and causes demyelination. *Cell Microbiol.* **2015**, *17*, 369–388. [CrossRef]
21. Rashid Rumah, K.; Eleso, O.; Fischetti, V.A.J.B. Human blood exposure to *Clostridium perfringens* epsilon toxin may shed light on erythrocyte fragility during active multiple sclerosis. *bioRivx* **2019**, 789123. [CrossRef]
22. Abou Faycal, C.; Gazzeri, S.; Eymin, B. A VEGF-A/SOX2/SRSF2 network controls VEGFR1 pre-mRNA alternative splicing in lung carcinoma cells. *Sci. Rep.* **2019**, *9*, 336. [CrossRef]
23. He, C.; Liu, B.; Wang, H.Y.; Wu, L.; Zhao, G.; Huang, C.; Liu, Y.; Shan, B.; Liu, L. Inhibition of SRPK1, a key splicing regulator, exhibits antitumor and chemotherapeutic-sensitizing effects on extranodal NK/T-cell lymphoma cells. *BMC Cancer* **2022**, *22*, 1100. [CrossRef]
24. Freedman, J.C.; McClane, B.A.; Uzal, F.A. New insights into *Clostridium perfringens* epsilon toxin activation and action on the brain during enterotoxemia. *Anaerobe* **2016**, *41*, 27–31. [CrossRef]
25. Babele, P.; Verma, S.; Kumar, R.B.; Bhagyawant, S.S.; Kamboj, D.V.; Alam, S.I. Elucidation of protein biomarkers in plasma and urine for epsilon toxin exposure in mouse model. *Anaerobe* **2019**, *59*, 76–91. [CrossRef]
26. Berget, S.M.; Moore, C.; Sharp, P.A. Spliced segments at the 5′ terminus of adenovirus 2 late mRNA. *Proc. Natl. Acad. Sci. USA* **1977**, *74*, 3171–3175. [CrossRef]

27. Chow, L.T.; Gelinas, R.E.; Broker, T.R.; Roberts, R.J. An amazing sequence arrangement at the 5′ ends of adenovirus 2 messenger RNA. *Cell* **1977**, *12*, 1–8. [CrossRef]
28. David, C.J.; Manley, J.L. Alternative pre-mRNA splicing regulation in cancer: Pathways and programs unhinged. *Genes. Dev.* **2010**, *24*, 2343–2364. [CrossRef]
29. Ma, X.; He, F. Advances in the study of SR protein family. *Genom. Proteom. Bioinform.* **2003**, *1*, 2–8. [CrossRef]
30. Lv, Y.; Zhang, W.; Zhao, J.; Sun, B.; Qi, Y.; Ji, H.; Chen, C.; Zhang, J.; Sheng, J.; Wang, T.; et al. SRSF1 inhibits autophagy through regulating Bcl-x splicing and interacting with PIK3C3 in lung cancer. *Signal Transduct. Target. Ther.* **2021**, *6*, 108. [CrossRef]
31. Du, J.X.; Luo, Y.H.; Zhang, S.J.; Wang, B.; Chen, C.; Zhu, G.Q.; Zhu, P.; Cai, C.Z.; Wan, J.L.; Cai, J.L.; et al. Splicing factor SRSF1 promotes breast cancer progression via oncogenic splice switching of PTPMT1. *J. Exp. Clin. Cancer Res.* **2021**, *40*, 171. [CrossRef]
32. Wan, L.; Lin, K.T.; Rahman, M.A.; Ishigami, Y.; Wang, Z.; Jensen, M.A.; Wilkinson, J.E.; Park, Y.; Tuveson, D.A.; Krainer, A.R. Splicing Factor SRSF1 Promotes Pancreatitis and KRASG12D-Mediated Pancreatic Cancer. *Cancer Discov.* **2023**, *13*, 1678–1695. [CrossRef]
33. Lance, A.; Druhan, L.J.; Vestal, C.G.; Steuerwald, N.M.; Hamilton, A.; Smith, M.; Price, A.; Tjaden, E.; Fox, A.N.; Avalos, B.R. Altered expression of CSF3R splice variants impacts signal response and is associated with SRSF2 mutations. *Leukemia* **2020**, *34*, 369–379. [CrossRef]
34. Gout, S.; Brambilla, E.; Boudria, A.; Drissi, R.; Lantuejoul, S.; Gazzeri, S.; Eymin, B. Abnormal expression of the pre-mRNA splicing regulators SRSF1, SRSF2, SRPK1 and SRPK2 in non small cell lung carcinoma. *PLoS ONE* **2012**, *7*, e46539. [CrossRef]
35. Luo, C.; Cheng, Y.; Liu, Y.; Chen, L.; Liu, L.; Wei, N.; Xie, Z.; Wu, W.; Feng, Y. SRSF2 Regulates Alternative Splicing to Drive Hepatocellular Carcinoma Development. *Cancer Res.* **2017**, *77*, 1168–1178. [CrossRef]
36. Li, K.; Wang, Z. Splicing factor SRSF2-centric gene regulation. *Int. J. Biol. Sci.* **2021**, *17*, 1708–1715. [CrossRef]
37. Wan, L.; Yu, W.; Shen, E.; Sun, W.; Liu, Y.; Kong, J.; Wu, Y.; Han, F.; Zhang, L.; Yu, T.; et al. SRSF6-regulated alternative splicing that promotes tumour progression offers a therapy target for colorectal cancer. *Gut* **2019**, *68*, 118–129. [CrossRef]
38. Zhang, F.; Wang, H.; Yu, J.; Yao, X.; Yang, S.; Li, W.; Xu, L.; Zhao, L. LncRNA CRNDE attenuates chemoresistance in gastric cancer via SRSF6-regulated alternative splicing of PICALM. *Mol. Cancer* **2021**, *20*, 6. [CrossRef]
39. Li, Y.; Xu, J.; Lu, Y.; Bian, H.; Yang, L.; Wu, H.; Zhang, X.; Zhang, B.; Xiong, M.; Chang, Y.; et al. DRAK2 aggravates nonalcoholic fatty liver disease progression through SRSF6-associated RNA alternative splicing. *Cell Metab.* **2021**, *33*, 2004–2020.e9. [CrossRef]
40. Makishima, H.; Visconte, V.; Sakaguchi, H.; Jankowska, A.M.; Kar, S.A.; Jerez, A.; Przychodzen, B.; Bupathi, M.; Guinta, K.; Afable, M.G.; et al. Mutations in the spliceosome machinery, a novel and ubiquitous pathway in leukemogenesis. *Blood* **2012**, *119*, 3203–3210. [CrossRef]
41. Cazzola, M.; Rossi, M.; Malcovati, L. Biologic and clinical significance of somatic mutations of SF3B1 in myeloid and lymphoid neoplasms. *Blood* **2013**, *121*, 260–269. [CrossRef]
42. Seo, J.Y.; Lee, K.O.; Kim, S.H.; Kim, K.; Jung, C.W.; Jang, J.H.; Kim, H.J. Clinical significance of SF3B1 mutations in Korean patients with myelodysplastic syndromes and myelodysplasia/myeloproliferative neoplasms with ring sideroblasts. *Ann. Hematol.* **2014**, *93*, 603–608. [CrossRef]
43. Bullock, N.; Oltean, S. The many faces of SRPK1. *J. Pathol.* **2017**, *241*, 437–440. [CrossRef]
44. Liu, H.; Gong, Z.; Li, K.; Zhang, Q.; Xu, Z.; Xu, Y. SRPK1/2 and PP1$\alpha$ exert opposite functions by modulating SRSF1-guided MKNK2 alternative splicing in colon adenocarcinoma. *J. Exp. Clin. Cancer Res.* **2021**, *40*, 75. [CrossRef]
45. Patel, M.; Sachidanandan, M.; Adnan, M. Serine arginine protein kinase 1 (SRPK1): A moonlighting protein with theranostic ability in cancer prevention. *Mol. Biol. Rep.* **2019**, *46*, 1487–1497. [CrossRef]
46. Geng, Z.; Huang, J.; Kang, L.; Gao, S.; Yuan, Y.; Li, Y.; Wang, J.; Xin, W.; Wang, J. *Clostridium perfringens* epsilon toxin binds to erythrocyte MAL receptors and triggers phosphatidylserine exposure. *J. Cell Mol. Med.* **2020**, *24*, 7341–7352. [CrossRef]
47. Zhang, N.; Jiang, N.; Zhang, K.; Zheng, L.; Zhang, D.; Sang, X.; Feng, Y.; Chen, R.; Yang, N.; Wang, X.; et al. Landscapes of Protein Posttranslational Modifications of African Trypanosoma Parasites. *iScience* **2020**, *23*, 101074. [CrossRef]
48. Livak, K.J.; Schmittgen, T.D. Analysis of relative gene expression data using real-time quantitative PCR and the $2^{-\Delta\Delta Ct}$ Method. *Methods* **2001**, *25*, 402–408. [CrossRef]

*toxins*

MDPI

*Article*

# Epsilon Toxin from *Clostridium perfringens* Induces the Generation of Extracellular Vesicles in HeLa Cells Overexpressing Myelin and Lymphocyte Protein

Jonatan Dorca-Arévalo [1,2,3,*], Antonio Santana-Ruiz [1], Benjamín Torrejón-Escribano [1,4], Mireia Martín-Satué [1,5] and Juan Blasi [1,2,3]

1 Department of Pathology and Experimental Therapeutics, Faculty of Medicine and Health Sciences-Campus Bellvitge, University of Barcelona, 08907 Barcelona, Spain; asalcala@gmail.com (A.S.-R.); torrejonbenja@ub.edu (B.T.-E.); martinsatue@ub.edu (M.M.-S.); blasi@ub.edu (J.B.)
2 Laboratory of Molecular and Cellular Neurobiology, Bellvitge Biomedical Research Institute (IDIBELL), 08907 L'Hospitalet de Llobregat, Spain
3 Institute of Neuroscience, Bellvitge Health Sciences Campus, University of Barcelona, Carrer de la Feixa Llarga, s/n, 08907 L'Hospitalet de Llobregat, Spain
4 Scientific and Technological Centers (CCiTUB), Bellvitge Campus, University of Barcelona, Carrer de la Feixa Llarga, s/n, 08907 L'Hospitalet de Llobregat, Spain
5 Oncobell Program, Institut d'Investigació Biomèdica de Bellvitge (IDIBELL), CIBERONC, 08908 L'Hospitalet de Llobregat, Spain
* Correspondence: jondorca@ub.edu

**Abstract:** Epsilon toxin (ETX) from *Clostridium perfringens* is a pore-forming toxin (PFT) that crosses the blood–brain barrier and binds to myelin structures. In in vitro assays, ETX causes oligodendrocyte impairment, subsequently leading to demyelination. In fact, ETX has been associated with triggering multiple sclerosis. Myelin and lymphocyte protein (MAL) is widely considered to be the receptor for ETX as its presence is crucial for the effects of ETX on the plasma membrane of host cells that involve pore formation, resulting in cell death. To overcome the pores formed by PFTs, some host cells produce extracellular vesicles (EVs) to reduce the amount of pores inserted into the plasma membrane. The formation of EVs has not been studied for ETX in host cells. Here, we generated a highly sensitive clone from HeLa cells overexpressing the MAL-GFP protein in the plasma membrane. We observed that ETX induces the formation of EVs. Moreover, the MAL protein and ETX oligomers are found in these EVs, which are a very useful tool to decipher and study the mode of action of ETX and characterize the mechanisms involved in the binding of ETX to its receptor.

**Keywords:** epsilon toxin; *Clostridium perfringens*; myelin and lymphocyte protein; extracellular vesicles

**Key Contribution:** The epsilon toxin induces the generation of extracellular vesicles in cells that express the MAL protein. In these vesicles, both the epsilon toxin and its receptor, the MAL protein, are detected, opening a new path of investigation into the mechanism of action of the epsilon toxin in host cells.

## 1. Introduction

*Clostridium perfringens*, a pore-forming Gram-positive anaerobic bacterium, is typically found in the gastrointestinal tract of mammals, including humans [1]. The biotyping of *C. perfringens* is performed using letters (A–G) to distinguish different toxin profiles. Among them, the epsilon toxin (ETX) is a pore-forming toxin (PFT), being the third most potent toxin after the tetanus and botulinum neurotoxins. Only types B and D produce the epsilon protoxin (pETX), which exhibits minimal activity but becomes fully active when it undergoes proteolytic cleavage [2].

ETX produces enterotoxemia in ruminants [3,4], causing multiorgan lesions and death. Furthermore, ETX can cross the blood–brain barrier (BBB) [5–7], potentially through vascular endothelial cell damage [4] or relaxation of the astrocyte endfeet terminals surrounding blood vessels [7–9]. Moreover, ETX produces neuronal damage in murine models [10,11]. Interestingly, ETX has been observed to preferentially accumulate in the kidneys and brain after an acute intravenous injection of ETX in mice, leading to brain edema and the death of epithelial cells in the distal tubules of the kidneys [12,13]. This effect was not observed in mice injected with the non-toxic ETX mutant H106P or pETX [14]. Notably, the Madin-Darby canine kidney (MDCK) cell line, derived from the distal convoluted tubule of dog kidneys, is particularly sensitive to the effects of ETX [15–18].

Furthermore, various cell lines of different origins show sensitivity to ETX, including Fischer rat thyroid (FRT) epithelial cells [19], MOLT-4 human T lymphoblast cells [20], and mouse lung endothelial 1G11 cells [21]. ETX has also been found to bind to and kill primary human lymphocytes [22] and produce hemolysis in human erythrocytes [23,24], likely through pore formation in the plasma membrane that is linked to the presence of myelin and lymphocyte protein (MAL). MAL is considered to be the ETX receptor in these sensitive cells [25], as recent research has suggested that the MAL protein is a potential receptor for ETX [26].

MAL is an integral membrane protein found in lipid rafts and plays a role in apical protein transport in polarized epithelial cells [27]. MAL is present in the myelin produced by oligodendrocytes and Schwann cells, as well as in the epithelium of the distal collecting duct of the kidneys, the urothelium, and the plasma membrane of T cells, coinciding with the areas where ETX binding is observed [26,28–30]. Moreover, overexpression of the MAL protein in different ETX-resistant cell lines lacking endogenous MAL increases their sensitivity to ETX, further supporting its role as an ETX receptor [26]. In addition, it has been reported that the knock-out of MAL gene expression in a T-lymphocyte cell line renders it resistant to ETX [20], while pulldown assays have confirmed a physical interaction between the MAL protein and ETX, providing additional evidence of MAL as the receptor for ETX [20,24].

Curiously, ETX has been associated with the triggering of the demyelinating disease multiple sclerosis (MS). *C. perfringens* types B and D have been detected in the stool samples from MS patients more frequently when compared to healthy control subjects [31,32]. Additionally, antibodies against ETX have been found in the serum of MS patients [32,33]. In vitro experiments have shown that ETX acts on oligodendrocytes [34–36], resulting in their death [36] and subsequently causing demyelination [35,36]. However, there is controversy regarding the mechanism of action of ETX in oligodendrocytes, with postulations including pore formation in the plasma membrane [36] or the activation of a pathway mediated by an undefined receptor that does not involve pore formation [35].

It is worth noting that some cells produce microvesicles (MVs) as a defense mechanism against various PFTs, such as streptolysin O, perfringolysin O, and pneumolysin [37–40], to reduce the presence of pores induced by these toxins in the plasma membrane. In addition, it has been recently published that ETX induces the generation of extracellular vesicles in *Xenopus* oocytes expressing human MAL protein [41]. In the present study, we utilized the HeLa cell line transfected with MAL-GFP to investigate more deeply the mechanism of action of ETX and the cell response. We show that ETX induces the budding of the plasma membrane, inducing extracellular vesicles (EV) formation, which is dependent on MAL. This opens a new field of study to decipher the mode of action of ETX.

## 2. Results

### 2.1. The Generated Clone of MAL-GFP-Expressing HeLa Cells Is Highly Sensitive to ETX

We performed cell subcloning from a population of stably transfected MAL-GFP-expressing HeLa cells [20] to obtain a pure and homogeneous cell population in order to study more accurately the effects of ETX on MAL-GFP-overexpressing HeLa cells and also to avoid the interference from non-MAL-expressing cells.

The selected MAL-GFP-overexpressing HeLa cell clone was incubated with the anti-MAL antibody E-1 to first corroborate MAL protein expression in the plasma membrane. While the MAL protein was not observed in the plasma membrane of the control GFP-overexpressing cells (Figure 1A–C), MAL protein immunostaining was positively detected in the MAL-GFP overexpressing cells (Figure 1D–F).

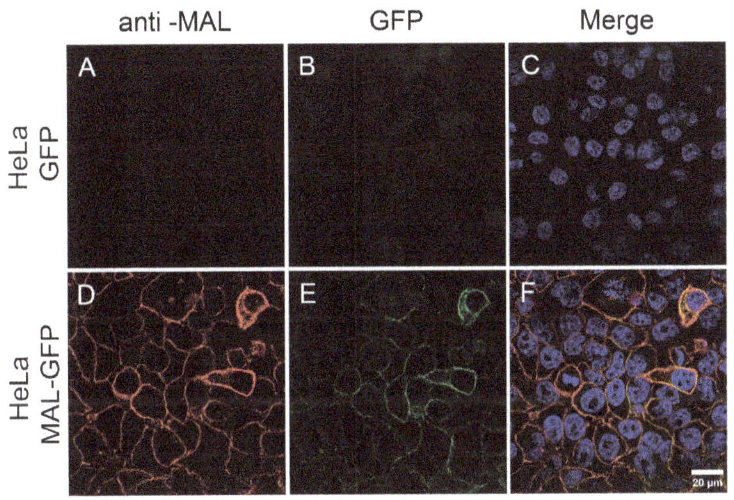


**Figure 1.** The MAL-GFP-expressing HeLa cell clone expresses high levels of the MAL protein in the plasma membrane. Cells were incubated with an anti-MAL antibody to detect MAL protein expression in the cell plasma membrane. The presence of MAL was compared between the GFP-expressing HeLa cells that did not express MAL (**A–C**) and the MAL-GFP-expressing HeLa cells (**D–F**). The MAL protein detected by the antibody is shown in red (**A**,**D**), while the MAL protein in fusion with GFP is shown in green (**B**,**E**). (**C**,**F**) The merge in each case. Note the complete colocalization between the antibody binding sites and the MAL-GFP protein distribution (**F**) as well as the absence of MAL expression in the HeLa cells expressing GFP alone (**C**). Nuclei were stained in blue.

Additionally, the MTS assays revealed that HeLa cells overexpressing MAL-GFP were highly sensitive to ETX (in red, Figure 2), with a $CT_{50}$ of 1.03 nM (95% CI, 0.87 to 1.22), thereby showing similar $CT_{50}$ values as MDCK cells [21,30]. No cytotoxicity was detected in the GFP-overexpressing HeLa cells (in black, Figure 2) incubated with ETX, confirming previous observations that MAL is crucial for ETX activity (Figure 2).

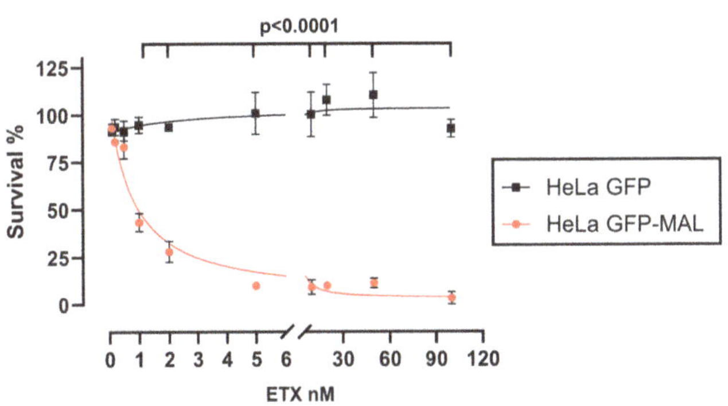


**Figure 2.** The MAL-GFP-expressing HeLa cell clone is highly sensitive to ETX. Cytotoxicity assays revealed a high sensitivity of the MAL-GFP-expressing HeLa cells (red line) to ETX, while no toxicity

was detected in the non-MAL-expressing HeLa cells that expressed GFP alone (black line). Cells were incubated with increasing concentrations of ETX (0, 0.25, 0.5, 1, 2, 5, 10, 25, 50, and 100 nM) for 1 h at 37 °C. Controls were obtained by omitting ETX in each condition (100% cell viability) or by adding 0.2% Triton X-100 (100% cell lethality). Triplicates of the assay were performed in three independent experiments for each condition. Statistical analyses were undertaken by performing a nonlinear regression analysis using a two-way ANOVA followed by Šídák's multiple comparisons test ($p < 0.0001$).

### 2.2. ETX Induces the Formation of EVs in MAL-GFP-Expressing HeLa Cells

Some cells produce MV in response to some PFTs [37,42,43]. As ETX belongs to the PFT family, we wondered if ETX also induced MV formation in host cells. As there are no studies on ETX and MV formation, we checked this possible effect by incubating MAL-GFP-overexpressing cells with ETX.

MAL-GFP-overexpressing HeLa cells were incubated with pETX-633 or ETX-633 for 40 min, and the effects of ETX on MV formation were analyzed. pETX-633 and ETX-633 bound to the cell plasma membrane at 5 min of incubation (in red, Figure 3, Videos S1 and S2). After 20 min of exposure, the binding of pETX-633 on the plasma membrane did not change and no MV were detected (B and E, Figure 3 and Video S1). Curiously, the binding pattern of ETX-633 on the plasma membrane was different, with a time-dependent decrease in ETX-633 binding (H and K, Figure 3 and Video S2) and the appearance of MV secreted by the cell (arrows in J and L, Figure 3 and Video S2).

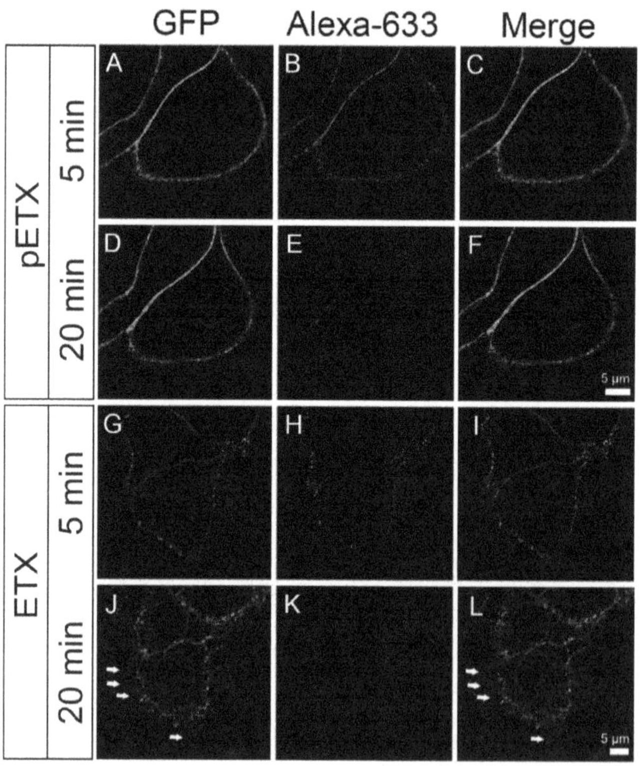


**Figure 3.** ETX induces the formation of extracellular vesicles in MAL-GFP-expressing HeLa cells. MAL-GFP-expressing HeLa cells (green) were incubated with 25 nM pETX-633 or ETX-633 (red) at

37 °C and under 5% $CO_2$ for 40 min. Images were obtained at 5 min and 20 min in each condition. pETX-633 bound to the plasma membrane of the cell, and the binding pattern did not change during the first 20 min of exposure (**A–F**). ETX-633 also bound to the plasma membrane (**G–L**), but after 20 min of exposure, the amount of bound ETX-633 decreased notably (**K**). In addition, the pattern of MAL-GFP staining changed over the time-course, and extracellular vesicles generated in response to ETX-633 were detected (arrows, **J**,**L**).

It can be observed that the ETX action generates MVs of varying sizes, consistent with the size range of EVs reported in the literature [44,45].

### *2.3. Characterization of the EVs Released in Response to ETX*

To characterize the production of EVs induced by ETX, the differential ultracentrifugation protocol was performed and the protein composition of the obtained EVs from each centrifugation step was analyzed by Western blot. We isolated the large EVs (LEVs) at 2000 g, the medium EVs (MEVs) at 10,000 g, and the small EVs (SEVs) at 150,000 g. The protein content-based EV characterization was conducted following the minimal information for studies of extracellular vesicles (MISEV) [45,46]. According to the guidelines, CD63, ITGB2, and LAMP1 were used as transmembrane markers, while ANXA2, TSG101, and HSP70 were used as markers of cytosolic proteins recovered from EVs. Calnexin (CANX) and ribosomal proteins (RPL30) were used as purity control markers.

pETX and ETX were detected at all the EV samples as a monomer (black and dark gray arrow, respectively, in Figure 4).

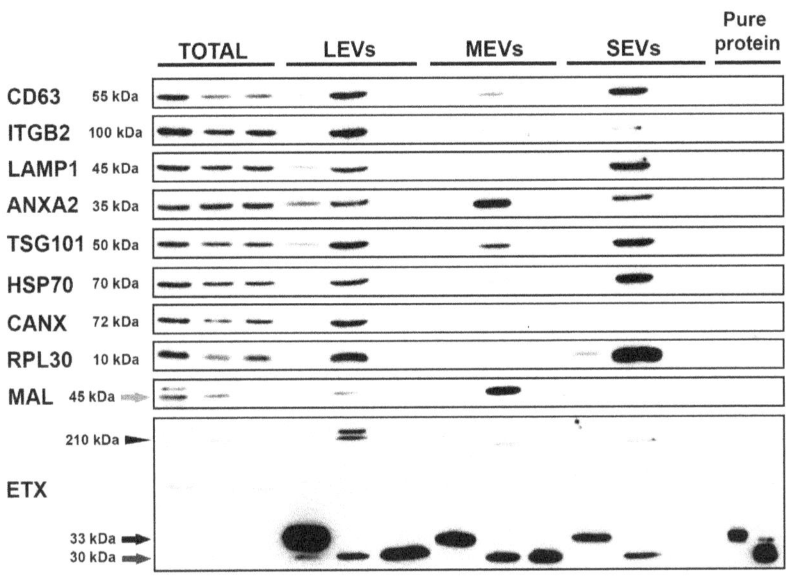


| HeLa GFP-MAL | + | + | − | + | + | − | + | + | − | + | + | − | − | − |
|---|---|---|---|---|---|---|---|---|---|---|---|---|---|---|
| HeLa GFP | − | − | + | − | − | + | − | − | + | − | − | + | − | − |
| pETX | + | − | − | + | − | − | + | − | − | + | − | − | + | − |
| ETX | − | + | + | − | + | + | − | + | + | − | + | + | − | + |

**Figure 4.** ETX induces the formation of extracellular vesicles with a high content of the MAL protein. Cell media were collected, and a differential ultracentrifugation protocol was applied to isolate the

different types of vesicles. ETX oligomer complexes (ETX heptamers) were detected in all the generated vesicles (arrowhead), as well as the monomer forms of both pETX and ETX (black and dark gray arrows, respectively). Moreover, the MAL protein was purified mostly from the MEVs compared to the total sample and the other vesicles (light gray arrow). (Total, total protein sample; LEVs, large extracellular vesicles; MEVs, medium extracellular vesicles; SEVs, small extracellular vesicles). The last two lanes correspond to the pure pETX and ETX proteins, respectively. Samples were run in two gels; total and LEVs were run in one gel, and MEVs, SEVs, and pure protein were run in another gel.

Moreover, ETX was also detected as an oligomer complex only in the MAL-expressing cells (black arrowhead, Figure 4). In addition, the MAL protein was mostly purified from the fraction obtained at 10,000 g, corresponding to MEVs. A faint band corresponding to the MAL protein was detected in the LEVs, while no band was observed in the SEVs. The ETX monomer was not detected in the SEVs obtained from HeLa cells expressing GFP alone, as expected.

pETX did not induce the secretion of any EVs in the MAL-GFP-expressing HeLa cells or in the GFP-expressing HeLa cells. However, ETX induced the formation of LEVs, MEVs, and SEVs only in the MAL-GFP-expressing HeLa cells. Curiously, the biochemical composition of the LEVs, MEVs, and SEVs was quite different, with the MEVs being the most different compared to the LEVs and SEVs. Transmembrane proteins were detected in all EVs, but in lower amounts in the MEVs compared to the LEVs and SEVs. Some of the cytosolic proteins recovered from the MEVs, such as ANXA2 and TSG101, were similar to those recovered from the LEVs and SEVs. CANX was only detected in the LEVs, which are very big and dense. RLP30 were detected only in the LEVs and SEVs, not in the MEVs.

## 3. Discussion

To study cell blebbing and EV formation in response to ETX, we generated a cell model that overexpresses the MAL protein and compared it to non-MAL-expressing cells. We showed that the generated clone expressed human MAL protein in the plasma membrane and that ETX bound to it and induced cell cytotoxicity. Furthermore, the clone presented a $CT_{50}$ value of 1.03 nM (95% CI, 0.87 to 1.22), which is quite similar to that of MDCK cells that show a $CT_{50}$ value of 0.64 nM (95% CI, 0.52 to 0.78) [20,21]. This clone was three times more sensitive to ETX than stably transfected MAL-GFP-expressing HeLa cells, which showed a $CT_{50}$ value of 3.36 nM (95% CI, 2.63 to 4.29) [20]. ETX produced blebbing in MAL-GFP-expressing HeLa cells after 20 min of exposure, and some EVs of different sizes were released from the plasma membrane (arrows, J and L in Figure 3 and Video S2). To analyze the composition of the EVs, a differential ultracentrifugation protocol using the media from cell cultures incubated with pETX or ETX was performed. While ETX induced the generation of EVs of different sizes, we ruled out the presence of apoptotic bodies since ETX did not elicit caspase-3 cleavage [41].

Some PFTs induce EV formation in order to eliminate the pores inserted into the plasma membrane [37,42,43]. The mechanisms by which these EVs mitigate damage are currently under investigation. It has been proposed that membrane repair occurs in two distinct steps: intrinsic repair and patch formation [37].

Intrinsic repair, which may function in an ATP- and protein-independent manner, refers to the lipid bilayer's inherent ability to resist PFTs through the biochemical and biophysical properties of membrane lipids. Lipid-modifying and lipid-binding enzymes, particularly sphingomyelinases, likely enhance and regulate this intrinsic repair process. Additionally, calcium influx through pores, combined with intrinsic repair, could initiate an intracellular response and trigger membrane shedding [42]. This calcium influx, promoted by the pore, could activate scramblase activity mediated by TMEM16F/ANO6 channels [41,47,48].

Certain repair proteins, such as Annexins and the ESCRT machinery, may also be recruited to sites of damage [49–53]. These proteins could play a critical role in sealing the damage and facilitating patch repair [54].

In line with these results, in *Xenopus* oocytes, ETX activates calcium-activated chloride channels (CaCCs) via intracellular $Ca^{2+}$ mobilization, inducing the activation of the scramblase and the formation of EVs that carry ETX heptamers [41]. Similarly, we speculate that EV formation in MAL-GFP-overexpressing HeLa cells could involve scramblases, a group of enzymes that facilitate the exchange of phospholipids between the two leaflets of the membrane [55]. However, further studies are needed to validate this hypothesis in mammalian cells.

Early studies on EVs were conducted in cancer research, where EVs were classified into three groups: apoptotic bodies, large extracellular vesicles, and exosomes [45,56]. Since the apoptotic pathway is not involved in the action of ETX [41], we adapted the terminology accordingly: large extracellular vesicles (LEVs), medium extracellular vesicles (MEVs), and small extracellular vesicles (SEVs).

We observed that ETX oligomerizes only if the MAL protein is expressed. An analysis of different types of EVs revealed that the MAL protein was found in some EVs following ETX exposure, with the MEVs (isolated at 10,000 g) being the main EV population in which the MAL protein was present. We propose that ETX binds to the MAL protein and oligomerizes, producing a pore that disrupts the plasma membrane integrity. This induces a rapid loss of intracellular $K^+$ ions and an increase in $Cl^-$ and $Na^+$ levels, subsequently eliciting a $Ca^{2+}$ influx [17] and TMEM16/ANO6 activation as well as the release of EVs containing ETX oligomers and MAL among other proteins. These effects are similar to those of the *Listeria* toxin [43]. Alternative binding partners for ETX, such as the O-glycosylated hepatitis A virus cellular receptor 1 (HAVCR1) protein, caveolin-1 (CAV1), and sulfatide, have been proposed, but they show lower binding efficiency [57–59]. These alternative binding partners could explain the presence of ETX oligomers at the SEVs in which the MAL protein was not detected as well as the detection of ETX monomers at the LEVs and MEVs from GFP-expressing HeLa cells that did not express the MAL protein. They could also explain the ability of ETX to oligomerize and form pores in synthetic lipid bilayers in which the MAL protein is not present [17,60], probably due to the presence of a required lipid environment. This occurs at a much lower capacity because ETX pores in planar lipid bilayers are formed by spontaneous self-assembly in the presence of 0.1–0.8 μM ETX [60], whereas MAL-GFP-expressing HeLa cells and MDCK cells show cytotoxicity in response to ETX at a $CT_{50}$ of 1.03 nM and 0.64 nM, respectively. Thus, the effect of ETX is a 100-fold more potent in mammalian cells compared to the synthetic lipid bilayers.

pETX induced the generation of a low amount of LEVs in MAL-GFP-expressing HeLa cells, as very faint bands corresponding to CD63, LAMP1, ANXA2, and TSG101 were detected. This could be explained by the fact that pETX is 1000-fold less toxic than ETX [61] but this toxicity is enough to activate the generation of EVs.

It is known that some PFTs induce the generation of EVs, which are produced by plasma membrane budding [43,62].

The origin of EVs secreted by plasma membrane budding (MEVs) differs from that of exosomes or smaller vesicles (SEVs) [63]. SEVs are enriched in some transmembrane proteins (CD63, ITGB2, and LAMP1) and RNA (RPL30), which aligns with typical exosomal markers originating from multivesicular bodies through an early endosome pathway [46,63].

In contrast, MEVs contain ANXA2 and, to a lesser extent, TSG101 (a component of the ESCRT-I complex). These proteins are believed to be recruited to sites of membrane damage, where they help seal and facilitate membrane repair in response to PFT action [49]. Furthermore, LEVs are the largest EVs generated by the cells and display all the tested markers, similar to the total homogenate, which may correspond to cell fragments, potentially associated with the necrotic pathway induced by ETX [64,65].

Altogether, ETX induces the generation of EVs from host cells, with the majority of the MEVs containing the MAL protein among others. This finding is very useful to understand, characterize, and decipher the mode of action of ETX in the presence or absence of certain proteins that are involved in this process.

## 4. Materials and Methods

### 4.1. Cell Lines

The epithelial cell line from a human cervical adenocarcinoma, HeLa (CCL-2 and ATCC), was selected as it does not express MAL and is not sensitive to ETX.

HeLa cells were grown as explained in [20] and cell cloning from the HeLa cells stably transfected with pEGFPN1-hMAL or pEGFPN1 was performed by applying the limiting dilution assay [66].

### 4.2. Cloning, Expression, Purification, and Activation of pETX and pETX-633

In all cases, wild-type DNA was cloned into the pGEX-4T-1 vector (Amersham Biosciences; Freiburg, Germany) to produce the corresponding recombinant fusion protein, as previously described [11,67,68]. Protein expression protocol was carried out as described in [14,29,30].

Some eluted pETX was labeled with DyLight™ 633 NHS Ester (#46414, Thermo Scientific, Waltham, MA, USA), following the manufacturer's instructions.

Purified pETX and pETX-633 were incubated with trypsin to form active ETX and ETX-633, respectively [69]. Preparations containing pETX and pETX-633 were incubated with trypsin-agarose beads (Sigma-Aldrich, St. Louis, MO, USA) at room temperature for 60 min, before the removal of the trypsin-coated beads by centrifugation. Protein concentration was determined following the Bradford method [70] using bovine serum albumin (BSA) as the standard.

In total, 10 µL of each sample was electrophoresed in a precast polyacrylamide gel (Mini-PROTEAN® TGX TM; #456-9033, Bio-Rad, Gaithersburg, MD, USA), and Coomassie blue staining was performed to analyze the purity of the recombinant protein.

### 4.3. Cytotoxicity Assays

The cytotoxic effect of ETX was measured using the MTS (3-(4,5- dimethylthiazol-2-yl)-5-(3-carboxymethoxyphenyl)-2-(4-sulfophenyl)-2H tetrazolium) colorimetric assay. Cells were exposed to increasing concentrations of ETX (0, 0.25, 0.5, 1, 2, 5, 10, 25, 50, and 100 nM) for 1 h at 37 °C, and the cytotoxicity assay was carried out as previously described [21]. The results from the cytotoxicity assays were analyzed by nonlinear regression analysis using a two-way ANOVA followed by Šídák's multiple comparisons test. $CT_{50}$ values were determined as explained previously [21].

### 4.4. Confocal Microscopy

Transfected HeLa cells were grown and fixed with 4% paraformaldehyde, and an immunofluorescence assay was performed as explained previously [20]. MAL protein was stained, incubating the cells with the anti-MAL E1 (Table 1), and was detected using Alexa Fluor 555 goat anti-mouse as a secondary antibody (1:500 dilution, A21424 Thermo Scientific, Waltham, MA, USA). Nuclei were stained with TO-PRO-3 iodide as explained previously [21].

**Table 1.** Table of antibodies used. Table of antibodies used. ANXA: anti-annexin A2; CANX: anti-calnexin; CD63: anti-CD63; ETX: anti-epsilon toxin; HSP70: anti-heat shock protein 70; ITGB2: anti-integrin β2; LAMP1: anti-lysosomal associated membrane protein 1; MAL: anti-myelin and lymphocyte protein; RLP30: anti-receptor like protein 30; TSG101: anti-tumor susceptibility gene 101.

| Antibody | Dilution | Host | Precedence |
|---|---|---|---|
| ANXA2 | 1/1000 | Rabbit | ab41803, Abcam |
| CANX | 1/1000 | Rabbit | ab133615, Abcam |
| CD63 | 1/1000 | Rabbit | ab134045, Abcam |
| ETX | 1/1000 | Rabbit | In-house |

**Table 1.** *Cont.*

| Antibody | Dilution | Host | Precedence |
|---|---|---|---|
| HSP70 | 1/1000 | Rabbit | ab181606, Abcam |
| ITGB2 | 1/1000 | Mouse | MBS9231826, abYntek |
| LAMP1 | 1/1000 | Rabbit | ab24170, Abcam |
| MAL | 1/300 | Mouse | SC-390687, Santacruz |
| RLP30 | 1/1000 | Rabbit | C119664, LSBio |
| TSG101 | 1/1000 | Rabbit | ab125011, Abcam |

The viability of the transfected HeLa cells was recorded in live-cell imaging time-course experiments, in which the cells were exposed to 25 nM ETX-633 (or pETX-633 for the negative control) for 40 min at 37 °C. The viability protocol was carried out as explained previously [21]. Images were recorded every 77 s for 40 min in each condition. pETX-633 or ETX-633 was added after the first 10 images were recorded as a baseline.

### 4.5. Microvesicle Isolation Protocol

Extracellular vesicles (EVs) were recovered from the different steps of the differential ultracentrifugation (DUC) protocol, which is classically used to isolate small extracellular vesicles or exosomes (SEVs), and the analysis of their protein composition was performed [47,48].

Briefly, cells grown in 10 cm plates were washed twice with PBS and incubated with 50 nM ETX, pETX, or PBS in DMEM-F12 serum-free media for 60 min at 37 °C and 5% $CO_2$. The media were collected and centrifuged at 300× *g* for 10 min at 4 °C. The resulting pellets were washed twice with cold PBS and kept on ice, while the supernatants were centrifuged at 2000× *g* for 10 min at 4 °C. The resulting pellets (large extracellular vesicles, LEVs) were washed twice with cold PBS and kept on ice, while the supernatants were centrifuged at 10,000× *g* for 35 min at 4 °C. The resulting pellets (medium extracellular vesicles, MEVs) were washed twice with cold PBS and kept on ice, while the supernatants were centrifuged at 150,000× *g* for 90 min at 4 °C. Finally, the resulting pellets (small extracellular vesicles, SEVs) were washed twice with cold PBS and kept on ice, while the supernatants were discarded. Samples were quantified by the Bradford assay. To characterize the EV composition, 80 μg of each sample supplemented with loading buffer was boiled and subjected to SDS-PAGE and Western blot analysis using the appropriate primary antibodies (Table 1).

**Supplementary Materials:** The following supporting information can be downloaded at https://www.mdpi.com/article/10.3390/toxins16120525/s1, Video S1: pETX-633 effect in HeLa MAL-GFP cells; Video S2: ETX-633 effect in HeLa MAL-GFP cells.

**Author Contributions:** Conceptualization, J.D.-A., J.B. and M.M.-S.; methodology, J.D.-A. and A.S.-R.; software, J.D.-A. and B.T.-E.; validation, J.D.-A., J.B. and M.M.-S.; formal analysis, J.D.-A., J.B. and M.M.-S.; investigation, J.D.-A. and J.B.; resources, J.D.-A. and J.B.; data curation, J.D.-A. and J.B.; writing—original draft preparation, J.D.-A.; writing—review and editing, J.D.-A., J.B. and M.M.-S.; visualization, J.D.-A. and J.B.; supervision, J.B.; project administration, J.B. and J.D.-A.; funding acquisition, J.B. and J.D.-A. All authors have read and agreed to the published version of the manuscript.

**Funding:** This study was supported by the Ministerio de Ciencia, Innovación y Universidades, Agencia Estatal de Investigación (AEI), through grants PID2021-126677NB-I00 funded by MCIN/AEI/10.13039/501100011033 and by the European Regional Development Fund (EDRF) "A way of making Europe", and the grant 21VAR008 funded by Neuroimpulse Program from the Institut d'Investigació Biomèdica de Bellvitge, IDIBELL. We thank the CERCA (Centres de Recerca de Catalunya, SGR: 2021 SGR 01075) Programme/Generalitat de Catalunya for institutional support.

**Institutional Review Board Statement:** Not applicable.

**Informed Consent Statement:** Not applicable.

**Data Availability Statement:** The original contributions presented in this study are included in this article and Supplementary Materials. Further inquiries can be directed to the corresponding author.

**Acknowledgments:** The authors thank the CCiTUB Bellvitge Campus Biology Unit for technical assistance. The authors also thank Inmaculada Gomez de Aranda for her technical support.

**Conflicts of Interest:** The authors declare no conflicts of interest.

## References

1. Stiles, B.G.; Barth, G.; Barth, H.; Popoff, M.R. *Clostridium perfringens* epsilon toxin: A malevolent molecule for animals and man? *Toxins* **2013**, *5*, 2138–2160. [CrossRef] [PubMed]
2. Bhown, A.S.; Habeerb, A.F. Structural studies on epsilon-prototoxin of *Clostridium perfringens* type D. Localization of the site of tryptic scission necessary for activation to epsilon-toxin. *Biochem. Biophys. Res. Commun.* **1977**, *78*, 889–896. [CrossRef] [PubMed]
3. Uzal, F.A.; Plumb, J.J.; Blackall, L.L.; O'Boyle, D.; Kelly, W.R. Detection by polymerase chain reaction of *Clostridium perfringens* producing epsilon toxin in faeces and in gastrointestinal contents of goats. *Lett. Appl. Microbiol.* **1996**, *23*, 13–17. [CrossRef]
4. Finnie, J.W.; Uzal, F.A. Pathology and Pathogenesis of Brain Lesions Produced by *Clostridium perfringens* Type D Epsilon Toxin. *Int. J. Mol. Sci.* **2022**, *23*, 9050. [CrossRef]
5. Gardner, D. Pathology of Clostridium-Welchii type-D enterotoxemia. 2. Structural and ultrastructural alterations in tissues of lambs and mice. *J. Comp. Pathol.* **1973**, *83*, 509–524. [CrossRef]
6. Morgan, K.; Kelly, B.; Buxton, D. Vascular leakage produced in brains of mice by Clostridium-welchii type-D toxin. *J. Comp. Pathol.* **1975**, *85*, 461–466. [CrossRef]
7. Finnie, J.W. Histopathological changes in the brain of mice given *Clostridium perfringens* type D epsilon toxin. *J. Comp. Pathol.* **1984**, *94*, 363–370. [CrossRef]
8. Finnie, J.W. Pathogenesis of brain damage produced in sheep by *Clostridium perfringens* type D epsilon toxin: A review. *Aust. Vet. J.* **2003**, *81*, 219–221. [CrossRef]
9. Freedman, J.C.; McClane, B.A.; Uzal, F.A. New insights into *Clostridium perfringens* epsilon toxin activation and action on the brain during enterotoxemia. *Anaerobe* **2016**, *41*, 27–31. [CrossRef]
10. Finnie, J.W.; Blumbergs, P.C.; Manavis, J. Neuronal damage produced in rat brains by *Clostridium perfringens* type D epsilon toxin. *J. Comp. Pathol.* **1999**, *120*, 415–420. [CrossRef]
11. Soler-Jover, A.; Dorca, J.; Popoff, M.R.; Gibert, M.; Saura, J.; Tusell, J.M.; Serratosa, J.; Blasi, J.; Martín-Satué, M. Distribution of *Clostridium perfringens* epsilon toxin in the brains of acutely intoxicated mice and its effect upon glial cells. *Toxicon* **2007**, *50*, 530–540. [CrossRef] [PubMed]
12. Tamai, E.; Ishida, T.; Miyata, S.; Matsushita, O.; Suda, H.; Kobayashi, S.; Sonobe, H.; Okabe, A. Accumulation of *Clostridium perfringens* epsilon-toxin in the mouse kidney and its possible biological significance. *Infect. Immun.* **2003**, *71*, 5371–5375. [CrossRef] [PubMed]
13. Soler-Jover, A.; Blasi, J.; Gómez de Aranda, I.; Navarro, P.; Gibert, M.; Popoff, M.R.; Martín-Satué, M. Effect of epsilon toxin-GFP on MDCK cells and renal tubules in vivo. *J. Histochem. Cytochem.* **2004**, *52*, 931–942. [CrossRef]
14. Dorca-Arevalo, J.; Pauillac, S.; Diaz-Hidalgo, L.; Martin-Satue, M.; Popoff, M.R.; Blasi, J. Correlation between In Vitro Cytotoxicity and In Vivo Lethal Activity in Mice of Epsilon Toxin Mutants from *Clostridium perfringens*. *PLoS ONE* **2014**, *9*, e102417. [CrossRef]
15. Payne, D.W.; Williamson, E.D.; Havard, H.; Modi, N.; Brown, J. Evaluation of a new cytotoxicity assay for *Clostridium perfringens* type D epsilon toxin. *FEMS Microbiol. Lett.* **1994**, *116*, 161–167. [CrossRef]
16. Petit, L.; Gibert, M.; Gillet, D.; Laurent-Winter, C.; Boquet, P.; Popoff, M.R. *Clostridium perfringens* epsilon-toxin acts on MDCK cells by forming a large membrane complex. *J. Bacteriol.* **1997**, *179*, 6480–6487. [CrossRef]
17. Petit, L.; Maier, E.; Gibert, M.; Popoff, M.R.; Benz, R. *Clostridium perfringens* epsilon toxin induces a rapid change of cell membrane permeability to ions and forms channels in artificial lipid bilayers. *J. Biol. Chem.* **2001**, *276*, 15736–15740. [CrossRef]
18. Petit, L.; Gibert, M.; Gourch, A.; Bens, M.; Vandewalle, A.; Popoff, M.R. *Clostridium perfringens* epsilon toxin rapidly decreases membrane barrier permeability of polarized MDCK cells. *Cell Microbiol.* **2003**, *5*, 155–164. [CrossRef]
19. Dorca-Arevalo, J.; Blanch, M.; Pradas, M.; Blasi, J. Epsilon toxin from *Clostridium perfringens* induces cytotoxicity in FRT thyroid epithelial cells. *Anaerobe* **2018**, *53*, 43–49. [CrossRef]
20. Blanch, M.; Dorca-Arevalo, J.; Not, A.; Cases, M.; de Aranda, I.; Martinez-Yelamos, A.; Martinez-Yelamos, S.; Solsona, C.; Blasi, J. The Cytotoxicity of Epsilon Toxin from *Clostridium perfringens* on Lymphocytes Is Mediated by MAL Protein Expression. *Mol. Cell. Biol.* **2018**, *38*, e00086-18. [CrossRef]
21. Dorca-Arevalo, J.; Dorca, E.; Torrejon-Escribano, B.; Blanch, M.; Martin-Satue, M.; Blasi, J. Lung endothelial cells are sensitive to epsilon toxin from *Clostridium perfringens*. *Vet. Res.* **2020**, *51*, 27. [CrossRef] [PubMed]
22. Shetty, S.V.; Mazzucco, M.R.; Winokur, P.; Haigh, S.V.; Rumah, K.R.; Fischetti, V.A.; Vartanian, T.; Linden, J.R. Epsilon Toxin Binds to and Kills Primary Human Lymphocytes. *Toxins* **2023**, *15*, 423. [CrossRef] [PubMed]
23. Gao, J.; Huang, J.; Ji, B.; Kang, L.; Gao, S.; Yang, H.; Shen, X.; Xin, W.; Wang, J. *Clostridium perfringens* epsilon-toxin induces human erythrocyte hemolysis by activation of the P2 receptor. *Toxicon* **2019**, *158*, S14. [CrossRef]
24. Geng, Z.; Huang, J.; Kang, L.; Gao, S.; Yuan, Y.; Li, Y.; Wang, J.; Xin, W. *Clostridium perfringens* epsilon toxin binds to erythrocyte MAL receptors and triggers phosphatidylserine exposure. *J. Cell Mol. Med.* **2020**, *24*, 7341–7352. [CrossRef] [PubMed]

25. Tilley, L.A.; Karamatic Crew, V.; Mankelow, T.J.; AlSubhi, S.A.; Jones, B.; Borowski, A.; Yahalom, V.; Finkel, L.; Singleton, B.K.; Walser, P.J.; et al. Deletions in the MAL gene result in loss of Mal protein, defining the rare inherited AnWj-negative blood group phenotype. *Blood* 2024, *Epub ahead of print*. [CrossRef]
26. Rumah, K.; Ma, Y.; Linden, J.; Oo, M.; Anrather, J.; Schaeren-Wiemers, N.; Alonso, M.; Fischetti, V.; McClain, M.; Vartanian, T. The Myelin and Lymphocyte Protein MAL Is Required for Binding and Activity of *Clostridium perfringens* epsilon-Toxin. *PLoS Pathogens* **2015**, *11*, e1004896. [CrossRef]
27. Lara-Lemus, R. On The Role of Myelin and Lymphocyte Protein (MAL) In Cancer: A Puzzle with Two Faces. *J. Cancer* **2019**, *10*, 2312–2318. [CrossRef]
28. Rubio-Ramos, A.; Labat-de-Hoz, L.; Correas, I.; Alonso, M.A. The MAL Protein, an Integral Component of Specialized Membranes, in Normal Cells and Cancer. *Cells* **2021**, *10*, 1065. [CrossRef]
29. Dorca-Arevalo, J.; Soler-Jover, A.; Gibert, M.; Popoff, M.; Martin-Satue, M.; Blasi, J. Binding of epsilon-toxin from *Clostridium perfringens* in the nervous system. *Vet. Microbiol.* **2008**, *131*, 14–25. [CrossRef]
30. Dorca-Arevalo, J.; Martin-Satue, M.; Blasi, J. Characterization of the high affinity binding of epsilon toxin from *Clostridium perfringens* to the renal system. *Vet. Microbiol.* **2012**, *157*, 179–189. [CrossRef]
31. Ma, Y.; Sannino, D.; Linden, J.; Haigh, S.; Zhao, B.; Grigg, J.; Zumbo, P.; Dündar, F.; Butler, D.; Profaci, C.; et al. Epsilon toxin-producing *Clostridium perfringens* colonize the multiple sclerosis gut microbiome overcoming CNS immune privilege. *J. Clin. Investig.* **2023**, *133*, e163239. [CrossRef] [PubMed]
32. Rumah, K.R.; Linden, J.; Fischetti, V.A.; Vartanian, T. Isolation of *Clostridium perfringens* type B in an individual at first clinical presentation of multiple sclerosis provides clues for environmental triggers of the disease. *PLoS ONE* **2013**, *8*, e76359. [CrossRef] [PubMed]
33. Wagley, S.; Bokori-Brown, M.; Morcrette, H.; Malaspina, A.; D'Arcy, C.; Gnanapavan, S.; Lewis, N.; Popoff, M.; Raciborska, D.; Nicholas, R.; et al. Evidence of *Clostridium perfringens* epsilon toxin associated with multiple sclerosis. *Mult. Scler. J.* **2019**, *25*, 653–660. [CrossRef] [PubMed]
34. Wioland, L.; Dupont, J.L.; Bossu, J.L.; Popoff, M.R.; Poulain, B. Attack of the nervous system by *Clostridium perfringens* Epsilon toxin: From disease to mode of action on neural cells. *Toxicon* **2013**, *75*, 122–135. [CrossRef]
35. Wioland, L.; Dupont, J.; Doussau, F.; Gaillard, S.; Heid, F.; Isope, P.; Pauillac, S.; Popoff, M.; Bossu, J.; Poulain, B. Epsilon toxin from *Clostridium perfringens* acts on oligodendrocytes without forming pores, and causes demyelination. *Cell. Microbiol.* **2015**, *17*, 369–388. [CrossRef]
36. Linden, J.R.; Ma, Y.; Zhao, B.; Harris, J.M.; Rumah, K.R.; Schaeren-Wiemers, N.; Vartanian, T. *Clostridium perfringens* Epsilon Toxin Causes Selective Death of Mature Oligodendrocytes and Central Nervous System Demyelination. *mBio* **2015**, *6*, e02513. [CrossRef]
37. Romero, M.; Keyel, M.; Shi, G.; Bhattacharjee, P.; Roth, R.; Heuser, J.E.; Keyel, P.A. Intrinsic repair protects cells from pore-forming toxins by microvesicle shedding. *Cell Death Differ.* **2017**, *24*, 798–808. [CrossRef]
38. Köffel, R.; Wolfmeier, H.; Larpin, Y.; Besançon, H.; Schoenauer, R.; Babiychuk, V.S.; Drücker, P.; Pabst, T.; Mitchell, T.J.; Babiychuk, E.B.; et al. Host-Derived Microvesicles Carrying Bacterial Pore-Forming Toxins Deliver Signals to Macrophages: A Novel Mechanism of Shaping Immune Responses. *Front. Immunol.* **2018**, *9*, 1688. [CrossRef]
39. Brito, C.; Cabanes, D.; Sarmento Mesquita, F.; Sousa, S. Mechanisms protecting host cells against bacterial pore-forming toxins. *Cell Mol. Life Sci.* **2019**, *76*, 1319–1339. [CrossRef]
40. Larpin, Y.; Besançon, H.; Iacovache, M.I.; Babiychuk, V.S.; Babiychuk, E.B.; Zuber, B.; Draeger, A.; Köffel, R. Bacterial pore-forming toxin pneumolysin: Cell membrane structure and microvesicle shedding capacity determines differential survival of cell types. *FASEB J.* **2020**, *34*, 1665–1678. [CrossRef]
41. Cases, M.; Dorca-Arévalo, J.; Blanch, M.; Rodil, S.; Terni, B.; Martín-Satué, M.; Llobet, A.; Blasi, J.; Solsona, C. The epsilon toxin from *Clostridium perfringens* stimulates calcium-activated chloride channels, generating extracellular vesicles in *Xenopus* oocytes. *Pharmacol. Res. Perspect.* **2024**, *12*, e70005. [CrossRef]
42. Keyel, P.A.; Loultcheva, L.; Roth, R.; Salter, R.D.; Watkins, S.C.; Yokoyama, W.M.; Heuser, J.E. Streptolysin O clearance through sequestration into blebs that bud passively from the plasma membrane. *J. Cell Sci.* **2011**, *124*, 2414–2423. [CrossRef] [PubMed]
43. Wu, N.; Cernysiov, V.; Davidson, D.; Song, H.; Tang, J.; Luo, S.; Lu, Y.; Qian, J.; Gyurova, I.E.; Waggoner, S.N.; et al. Critical Role of Lipid Scramblase TMEM16F in Phosphatidylserine Exposure and Repair of Plasma Membrane after Pore Formation. *Cell Rep.* **2020**, *30*, 1129–1140.e1125. [CrossRef] [PubMed]
44. Salem, I.; Naranjo, N.M.; Singh, A.; DeRita, R.; Krishn, S.R.; Sirman, L.S.; Quaglia, F.; Duffy, A.; Bowler, N.; Sayeed, A.; et al. Methods for extracellular vesicle isolation from cancer cells. *Cancer Drug Resist.* **2020**, *3*, 371–384. [CrossRef] [PubMed]
45. Théry, C.; Witwer, K.W.; Aikawa, E.; Alcaraz, M.J.; Anderson, J.D.; Andriantsitohaina, R.; Antoniou, A.; Arab, T.; Archer, F.; Atkin-Smith, G.K.; et al. Minimal information for studies of extracellular vesicles 2018 (MISEV2018): A position statement of the International Society for Extracellular Vesicles and update of the MISEV2014 guidelines. *J. Extracell. Vesicles* **2018**, *7*, 1535750. [CrossRef] [PubMed]
46. Welsh, J.A.; Goberdhan, D.C.I.; O'Driscoll, L.; Buzas, E.I.; Blenkiron, C.; Bussolati, B.; Cai, H.; Di Vizio, D.; Driedonks, T.A.P.; Erdbrügger, U.; et al. Minimal information for studies of extracellular vesicles (MISEV2023): From basic to advanced approaches. *J. Extracell. Vesicles* **2024**, *13*, e12404. [CrossRef] [PubMed]

47. Kowal, J.; Arras, G.; Colombo, M.; Jouve, M.; Morath, J.P.; Primdal-Bengtson, B.; Dingli, F.; Loew, D.; Tkach, M.; Théry, C. Proteomic comparison defines novel markers to characterize heterogeneous populations of extracellular vesicle subtypes. *Proc. Natl. Acad. Sci. USA* **2016**, *113*, E968–E977. [CrossRef]
48. Théry, C.; Amigorena, S.; Raposo, G.; Clayton, A. Isolation and characterization of exosomes from cell culture supernatants and biological fluids. *Curr. Protoc. Cell Biol.* **2006**, *30*, 3.22.1–3.22.29. [CrossRef]
49. Cooper, S.T.; McNeil, P.L. Membrane Repair: Mechanisms and Pathophysiology. *Physiol. Rev.* **2015**, *95*, 1205–1240. [CrossRef]
50. Demonbreun, A.R.; Quattrocelli, M.; Barefield, D.Y.; Allen, M.V.; Swanson, K.E.; McNally, E.M. An actin-dependent annexin complex mediates plasma membrane repair in muscle. *J. Cell Biol.* **2016**, *213*, 705–718. [CrossRef]
51. Babiychuk, E.B.; Monastyrskaya, K.; Potez, S.; Draeger, A. Blebbing confers resistance against cell lysis. *Cell Death Differ.* **2011**, *18*, 80–89. [CrossRef] [PubMed]
52. Jimenez, A.J.; Maiuri, P.; Lafaurie-Janvore, J.; Divoux, S.; Piel, M.; Perez, F. ESCRT machinery is required for plasma membrane repair. *Science* **2014**, *343*, 1247136. [CrossRef] [PubMed]
53. Jimenez, A.; Perez, F. Plasma membrane repair: The adaptable cell life-insurance. *Curr. Opin. Cell Biol.* **2017**, *47*, 99–107. [CrossRef] [PubMed]
54. McNeil, A.K.; Rescher, U.; Gerke, V.; McNeil, P.L. Requirement for annexin A1 in plasma membrane repair. *J. Biol. Chem.* **2006**, *281*, 35202–35207. [CrossRef]
55. Whitlock, J.M.; Hartzell, H.C. Anoctamins/TMEM16 Proteins: Chloride Channels Flirting with Lipids and Extracellular Vesicles. *Annu. Rev. Physiol.* **2017**, *79*, 119–143. [CrossRef]
56. Yáñez-Mó, M.; Siljander, P.R.; Andreu, Z.; Zavec, A.B.; Borràs, F.E.; Buzas, E.I.; Buzas, K.; Casal, E.; Cappello, F.; Carvalho, J.; et al. Biological properties of extracellular vesicles and their physiological functions. *J. Extracell. Vesicles* **2015**, *4*, 27066. [CrossRef]
57. Ivie, S.E.; Fennessey, C.M.; Sheng, J.; Rubin, D.H.; McClain, M.S. Gene-trap mutagenesis identifies mammalian genes contributing to intoxication by *Clostridium perfringens* epsilon-toxin. *PLoS ONE* **2011**, *6*, e17787. [CrossRef]
58. Adler, D.; Linden, J.; Shetty, S.; Ma, Y.; Bokori-Brown, M.; Titball, R.; Vartanian, T. *Clostridium perfringens* Epsilon Toxin Compromises the Blood-Brain Barrier in a Humanized Zebrafish Model. *iScience* **2019**, *15*, 39–54. [CrossRef]
59. Gil, C.; Dorca-Arevalo, J.; Blasi, J. *Clostridium perfringens* Epsilon Toxin Binds to Membrane Lipids and Its Cytotoxic Action Depends on Sulfatide. *PLoS ONE* **2015**, *10*, e0140321. [CrossRef]
60. Marshall, S.; McGill, B.; Morcrette, H.; Winlove, C.P.; Chimerel, C.; Petrov, P.G.; Bokori-Brown, M. Interaction of *Clostridium perfringens* Epsilon Toxin with the Plasma Membrane: The Role of Amino Acids Y42, Y43 and H162. *Toxins* **2022**, *14*, 757. [CrossRef]
61. Worthington, R.; Mulders, M. Physical Changes in Epsilon proto-toxin Molecule of *Clostridium perfringens* During Enzymatic Activation. *Infect. Immun.* **1977**, *18*, 549–551. [CrossRef] [PubMed]
62. Colombo, M.; Raposo, G.; Théry, C. Biogenesis, secretion, and intercellular interactions of exosomes and other extracellular vesicles. *Annu. Rev. Cell Dev. Biol.* **2014**, *30*, 255–289. [CrossRef] [PubMed]
63. Doyle, L.M.; Wang, M.Z. Overview of Extracellular Vesicles, Their Origin, Composition, Purpose, and Methods for Exosome Isolation and Analysis. *Cells* **2019**, *8*, 727. [CrossRef] [PubMed]
64. Geng, Z.; Kang, L.; Huang, J.; Gao, S.; Wang, J.; Yuan, Y.; Li, Y.; Xin, W. Epsilon toxin from *Clostridium perfringens* induces toxic effects on skin tissues and HaCaT and human epidermal keratinocytes. *Toxicon* **2021**, *198*, 102–110. [CrossRef]
65. Chehreara, R.; Karizi, S.Z.; Hosseini, H.M.; Mirhosseini, S.A.; Shafiei, M.; Amani, J.; Kazemi, R. Cell death analysis of recombinant mature epsilon toxin on the kidney cell line. *Iran. J. Microbiol.* **2021**, *13*, 832–838. [CrossRef]
66. Freshney, R.I. (Ed.) *Culture of Animal Cells: A Manual of Basic Technique and Specialized Applications*, 6th ed.; Wiley-Blackwell: Hoboken, NJ, USA, 2010; pp. 208–211.
67. Hunter, S.E.; Clarke, I.N.; Kelly, D.C.; Titball, R.W. Cloning and nucleotide sequencing of the *Clostridium perfringens* epsilon-toxin gene and its expression in Escherichia coli. *Infect. Immun.* **1992**, *60*, 102–110. [CrossRef]
68. Dorca-Arévalo, J.; Gómez de Aranda, I.; Blasi, J. New Mutants of Epsilon Toxin from *Clostridium perfringens* with an Altered Receptor-Binding Site and Cell-Type Specificity. *Toxins* **2022**, *14*, 288. [CrossRef]
69. Oyston, P.; Payne, D.; Havard, H.; Williamson, E.; Titball, R. Production of a non-toxic site-directed mutant of *Clostridium perfringens* epsilon-toxin which induces protective immunity in mice. *Microbiology* **1998**, *144*, 333–341. [CrossRef]
70. Bradford, M.M. A rapid and sensitive method for the quantitation of microgram quantities of protein utilizing the principle of protein-dye binding. *Anal. Biochem.* **1976**, *72*, 248–254. [CrossRef]

MDPI AG
Grosspeteranlage 5
4052 Basel
Switzerland
Tel.: +41 61 683 77 34

*Toxins* Editorial Office
E-mail: toxins@mdpi.com
www.mdpi.com/journal/toxins

www.ingramcontent.com/pod-product-compliance
Lightning Source LLC
Chambersburg PA
CBHW041953260926
51850CB00019B/2704
* 9 7 8 3 7 2 5 8 7 0 8 0 6 *